Evolved Art

Turtles – Volume One

By Tim Endres

To Sky Marie and The Boys

Version 1.8

First Printing June 2009
ISBN-10: 0-615-30034-0
ISBN-13: 978-0-615-30034-4
EvolvedArt.biz

Table of Contents

Chapter 1

Evolution

In 1981, I was attending General Motors Institute (GMI) in Flint, Michigan pursuing a Bachelors of Science in Electrical Engineering. I was a sophomore at the time. GMI was a small college fully subsidized by General Motors, which used the school to produce engineers and managers for the company's future. GMI was a cooperative program. You spent three months at school, then three months working for your sponsor. My sponsor was Oldsmobile of Lansing, Michigan.

While working for Oldsmobile during my sophomore year, I was exposed to a programming language named APL, which stood for A Programming Language. To this day, APL stands alone in terms of its uniqueness of expression and applicability to certain programming domains. APL was designed by Ken Iverson, who at the time was an IBM fellow. He developed the language at IBM, and I was programming with it on an IBM mainframe. I was working with someone in administration who was using a spreadsheet much like Lotus 123 to manage obsolescence at the company. Since the spreadsheet application was written in APL, and APL is an interpretive language, I was able to escape out of the spreadsheet into the interpreter and start writing my own programs.

One of the things that made APL so unique was that it was designed to be capable of directly programming mathematical expressions, especially those involving matrices. So unique was the language that if you coded a loop in your application, a fundamental construct of nearly every program ever written, you were using the language incorrectly!

Another feature of APL was that you could write APL expressions in your program and then have them interpreted. Very few languages at the time offered this capability. Since APL was designed to express mathematical equations, I came up with the idea of writing a program that would generate APL expressions and evaluate them for correctness in an attempt at automatically creating mathematical algorithms and theorems. Little did I realize I was blindly stumbling into an area that was beginning to show some fascinating potential – genetic programming.

Years after my feeble attempts to generate mathematical expressions with APL, I discovered genetic algorithms when I came upon the book *Adaptation in Natural and Artificial Systems* by John Holland, who popularized genetic algorithms in the early 1970's while at the University of Michigan. While I experimented with genetic algorithms, I discovered a related area of research called genetic programming when I read *Genetic Programming* by John Koza, published in 1992. Genetic programming is another method of evolving solutions, but brings some fundamental changes that I think make it a much more generalized system. A discussion of genetic algorithms and genetic programming is beyond the scope of this book, but there is an extensive body of material if you wish to further study these topics.

My early experiments with genetic programming, interestingly enough, were attempts to optimize searching. Google had never been heard of. In fact, the expression "www" had yet to be uttered. Two friends and I had founded Msen.com, an internet service provider, in 1991. We had a dedicated 56Kb circuit, and a contract with Reuters for exclusive distribution of Reuters News over the U.S. Internet. We had a satellite dish on the roof and I wrote a perl script to separate the incoming news feed into USENET news groups. There were no good tools for searching that news feed, and I had high hopes of finding a new technique.

Those early efforts were not successful. But I was fascinated with genetic programming, and would continue to experiment with it for many years. Some time before this, I was in Borders bookstore in Ann Arbor (yes, the original) and I purchased a book about Turtle Logo.

Turtle Logo is a programming language that is often used to teach programming and computer graphics concepts to beginners. The strength of Turtle Logo is its simplicity. The turtle is an imaginary creature whose tail is a pen. The turtle only knows how to move forward, turn right, turn left, and raise and lower its pen. However, using these simple commands, one can program some very nifty drawings.

Years later that book would surface while working on some genetic programming systems, and prompt an idea. As I was reading the book about Turtle Logo, a thought suddenly came to me. These turtle programs looked a little like the ones I was *evolving* using genetic programming. What if I wrote an application that would evolve Turtle Logo programs for me? Might I be able to coax these turtles into drawing some interesting graphics?

Like many ideas, this one stuck in my head for some time, but I never had the free time to work on it. That is, until in 1993 when I finally found enough time to code the application on my Apple Macintosh. On a small black-and-white screen that could only display a population of 24 turtles, I got my first glimpse at the potential of evolving art in a way that I believe had never been done before.

Years passed, and my work moved me off of the Macintosh and onto a Windows machine. Once again I had some free time and I was again inspired by the turtles. So in 1996, I endeavored to rewrite the turtle application for the Windows platform. I gave the application a name. I called it BioLogo.

When I saw the color turtles drawing on my computer screen for the first time, I must admit that I was completely blown away. In my wildest dreams, I did not imagine some of the images that would ultimately evolve. Stunning, complex, and beautiful, the turtles were a stark testimony to the power of genetic programming and evolutionary concepts. This was definitely a case of the creation surprising the creator.

I have since developed the application to be a capable Windows application, and have also developed a website that allows turtle evolvers to upload, organize, present, and share their favorite turtles. The site also provides forums for the discussion of turtles and evolved art in general. You can find the site at www.evolvedturtle.com.

Chapter 2

BioLogo

In this chapter I will discuss in very general terms the design of the BioLogo application and its underlying language. This is for those who might like a little better understanding of thc inner workings of the turtles and how they are evolved.

Turtle Logo is a very nice language with a serious following. It is a fairly complete language in that it provides for mathematical expressions, conditionals, loops, and functions. Its primary difference from other languages is its primary goal of drawing pictures. However, if one wished, one could write more general algorithms in Turtle Logo. That said, it would not likely be a first choice for anything other than graphics.

As a simple example, and to give you a taste for the language, here is a Turtle Logo program that will draw a simple square. This code tells the turtle to move forward 100 steps, turn right 90 degrees, and repeat the process a total of four times.

```
REPEAT 4 [ FORWARD 100 RIGHT 90 ]
```

My original idea was to develop an application that would evolve Turtle Logo programs using the entirety of the language. This had many advantages, including being able to load up existing Turtle Logo programs. However, when I set out to code my application, I began to see some issues with the original concept. I would either have to devote a great deal more time to the implementation of the evolutionary engine, or I was going to have to greatly simplify the language that I was evolving.

Since this is my hobby, I did not have the luxury of devoting the many hours it would have taken to support the complete Turtle Logo language. So I was forced to reduce the language to something that was optimized for genetic programming. In the end, Turtle Logo programmers will argue that my language is not even close to Turtle Logo, and they are correct. It really only retains the most core concepts of the drawing aspects of the language, and even then with a completely different syntax.

For comparison, here is a BioLogo program to draw the same simple square.

```
( REPEAT_2 ( REPEAT_2 ( ( FWD_100 ) ( RIGHT_90 ) ) ) )
```

As you can see, the code is quite different. I will not bore you with a full discussion of the language and the differences between BioLogo and Turtle Logo. Suffice it to say that I would never be offended by a Turtle Logo aficionado insisting that BioLogo has no business passing itself off as a relative of Turtle Logo.

The strength of the BioLogo language, and the intent of its design, is that you can take any BioLogo program, which is represented internally as a tree of nodes, break it apart at any node, and splice either of the two resulting subtree fragments with any other subtree fragment obtained by splitting any other BioLogo program, and the resulting BioLogo program is valid. This is very important. It means that I do not have to write code to ensure the validity of the recombination of any two nodes during crossover. Nor do I have write code to validate the overall program resulting from the recombination.

Take any two BioLogo programs, split them both at any arbitrary node, crossover the resulting code fragment subtrees to recombine into two new BioLogo programs, and the resulting programs are guaranteed to be valid. By valid we mean that the program will run without errors. We make no assumption about the resulting graphics that will be generated by running the program. The evolver (or BioLogo user) will perform the natural selection of desirable turtles.

Chapter 3

Future Directions

BioLogo, and its website are the culmination of over 15 years of hobby. I have built everything that I originally envisioned. However, there are still a few ideas I would like to pursue if I ever find the time.

There are several enhancements that I would like to make to the existing BioLogo application. For instance, I would really like to build an engine that implemented multiple turtles drawing simultaneously to produce a single image. This would allow for some interesting experiments coding the evolutionary engine, and I believe would allow for the drawing of much more complex and interesting images. I would also like to build analysis tools, such as an editor that will highlight the code that draws a part of the image that we highlight with the mouse.

I also hope to have time to make some enhancements to the website to improve the user experience as well and improve the ability to present and exchange turtles. In time, as computing power increases sufficiently, I would even like to provide the ability to evolve turtles on the website.

Another concept that intrigues me is the idea of implementing a turtle based on surface patches instead of lines. When I was at Oldsmobile, we were using surface patching to analyze car parts and the stresses on them. The part was broken down into small triangular patches that could be analyzed individually, then combined to present an overall analysis. Using these patches, we could build virtually any part we needed to analyze. The technique was called finite element analysis.

Like the finite element analysis, I imagine a triangular patch as the turtle. The patch would fill itself with a color, then it would draw triangular patches on each of its three sides, those triangles being determined by the program. There are numerous ways to implement this within the evolutionary engine. What I envision would be evolving surface drawings instead of line drawings.

Of course, one could take the concept a step, and dimension, further. With ray tracing capabilities on modern hardware, one could imagine a three dimensional turtle language. This is something I have not devoted much thought to, but it seems to have some potential.

There are also real world applications in nano-technology and biotechnology where fitness functions could allow genetic programming of structural and functional designs. One example is a group that is using proteins to construct objects. These proteins come in a set of simple shapes, and they are combined to create functional units, somewhat like playing with building blocks or legos. With a proper fitness function, we could conceivably have engineers evolve structures to solve problems.

Chapter 4

Turtles

While I have devoted some discussion to the inner workings of BioLogo, and the evolution of turtles, the real purpose of this book is to highlight some of the turtles that I have evolved over the years. In this chapter, I will display some of the thousands of turtles that I have deemed worth adding to my collection. You can also view my portfolios on the turtle website.

I first present a number of individual turtles that represent not only some of the more interesting turtles I have collected over the years, but also turtles that help illustrate the potential for complexity and diversity in what can be evolved.

Many times I have heard comments such as "oh, that reminds me of those old spirograph drawings", or "nice sine wave", or "did you make that with Photoshop?" While any of these comments are quite reasonable given any specific turtle, and I admit it is fun to evolve turtles that bring back memories of those old spirograph days, what each of these comments fails to recognize is the open-ended nature of the BioLogo system.

What I mean to say is that in theory you could evolve a turtle to draw any shape you desired. Given enough time, computing power, and determination, you could evolve a turtle that would draw the Empire State Building, or draw your signature! I have evolved turtles that draw perfect squares or nearly perfect circles with not too much effort. I hope that the turtles I present here can bring this point across.

In chapter five I present groups of four turtles on each page. These pages are showing turtles that are all related. Often times you can see the family resemblance, but there are times when the lineage is not at all obvious. This quality helps make evolving turtles exciting at times. You will be pursuing a lineage that seems promising, and suddenly a completely new idea will arise in the population that seems quite improbable.

I have been evolving and collecting turtles for over fifteen years, and I still feel a rush when I see something unique. I am continually impressed by the ability of the turtles to find unique drawings.

Let me start out with some of the most spiro-retro turtles that I can glean from the collection, after which I will slowly move to more unique shapes and symmetries to try and provide a sense of the ability of the turtles to move in directions that are not immediately obvious or necessarily intuitive. Note that there is a fitness bias for turtles to draw inside of a square, which favors circular drawings. It is entirely possible to parameterize the fitness formula to evolve turtles that are not at all circular, if one wished.

What I cannot demonstrate on these pages is what you will see when you watch BioLogo draw a turtle. Since the graphics take a small amount of time to actually draw many of the turtles' pictures, you can actually see the lines drawn out by the turtle. Many times, what appears to be a complex drawing is really just a simple figure that is repeated in ways that create the seeming complexity. Much like a good deal of nature.

Finally, please note that I present these turtles as-is. Just like human beings, turtles will occasionally have little moles. In other words, a turtle might draw a perfect mandala, then run to the middle of it and scribble a little in a different color. That little scribble can stick out and upset an otherwise perfect drawing. If I were creating large prints for framing, I might spend the time to remove these imperfections, but for this volume I present the turtles exactly as they evolved.

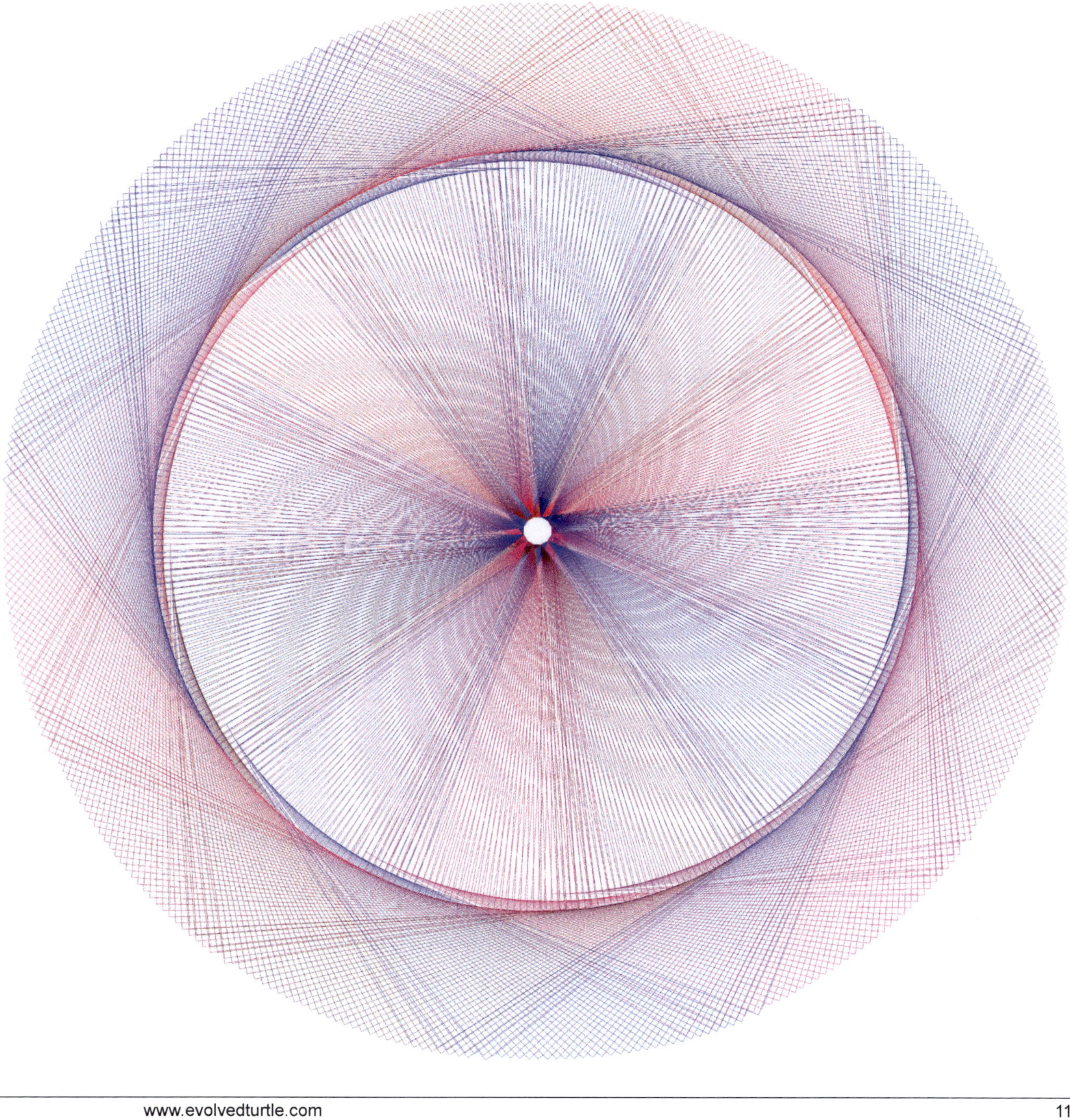

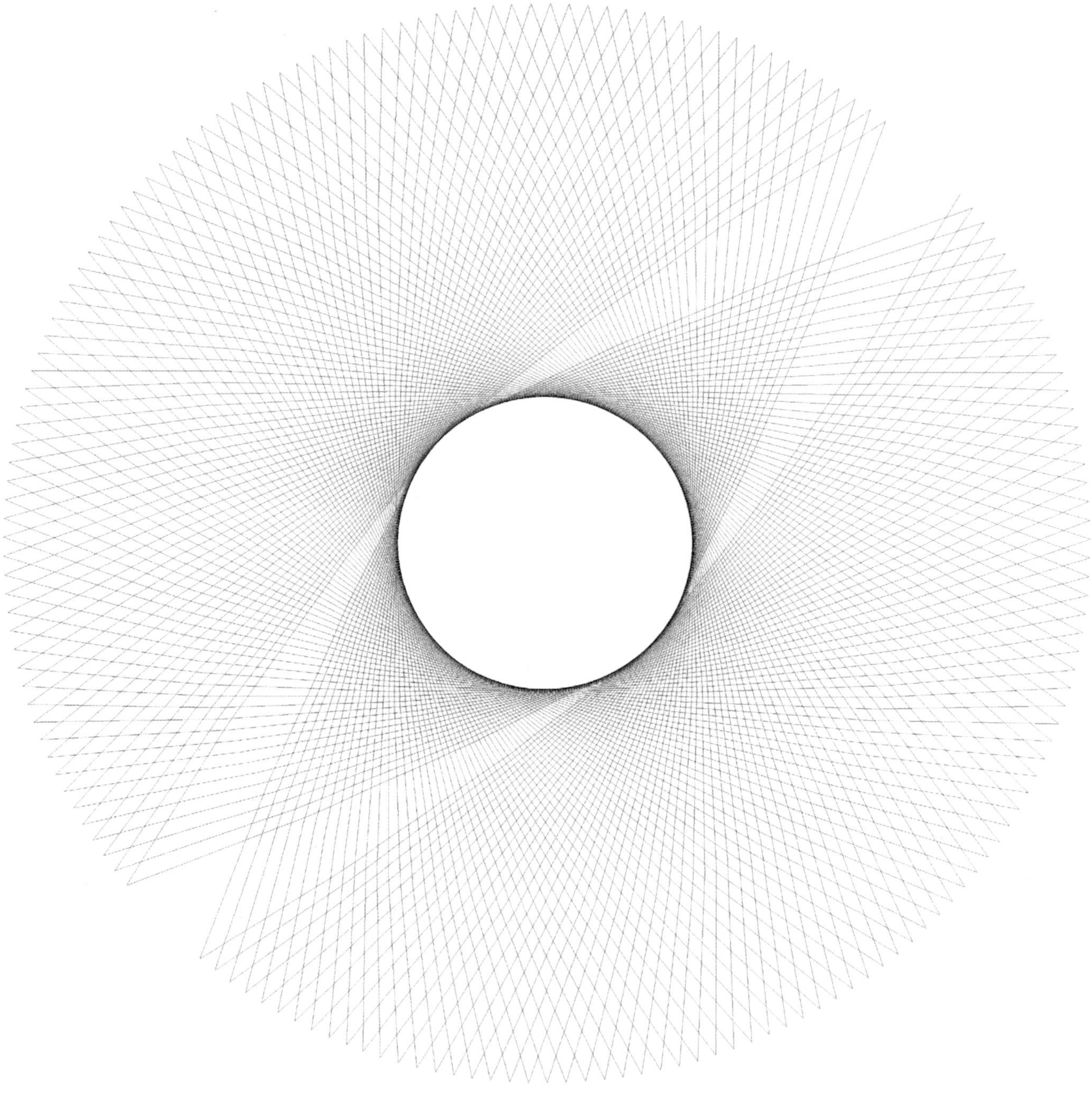

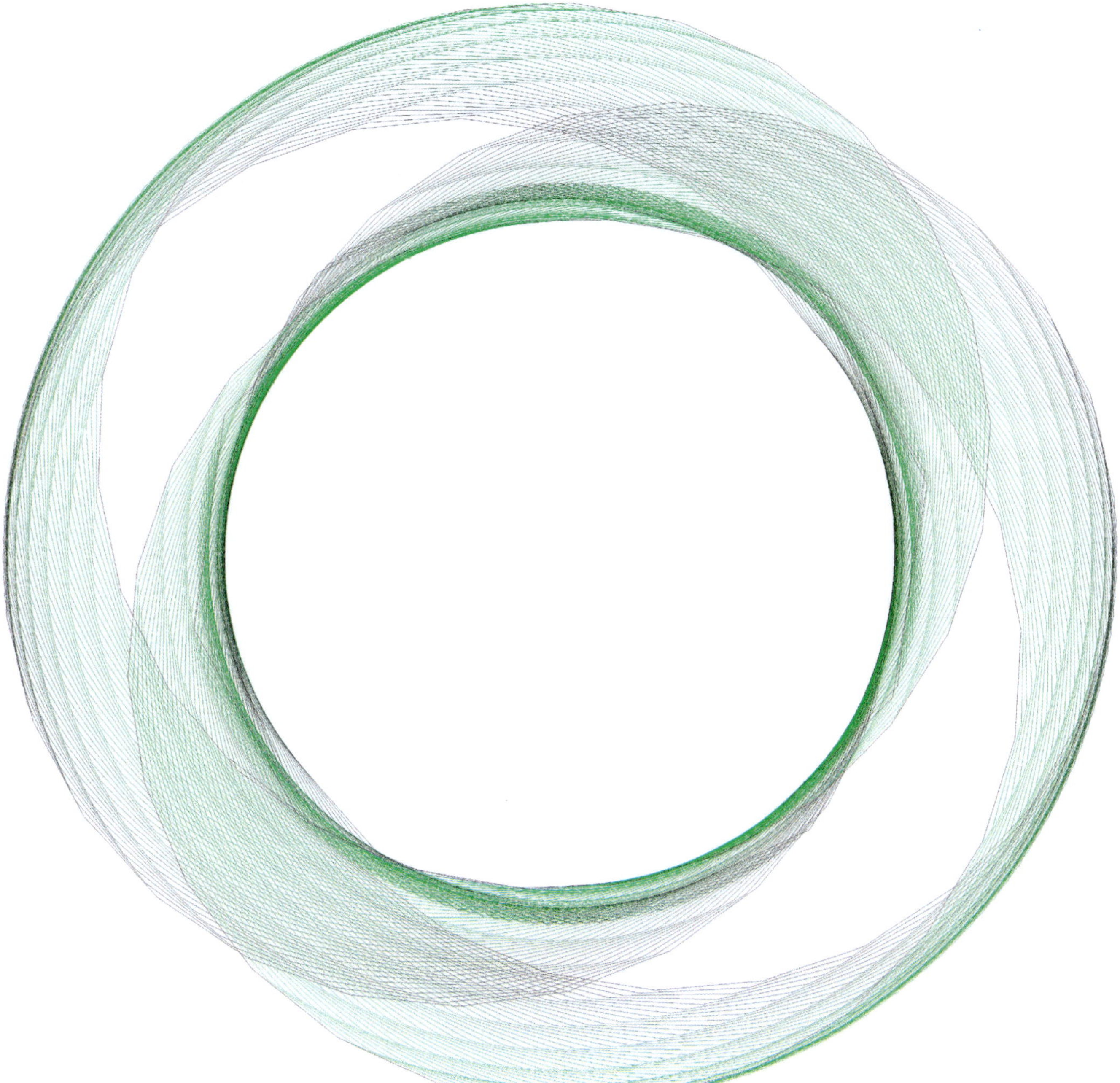

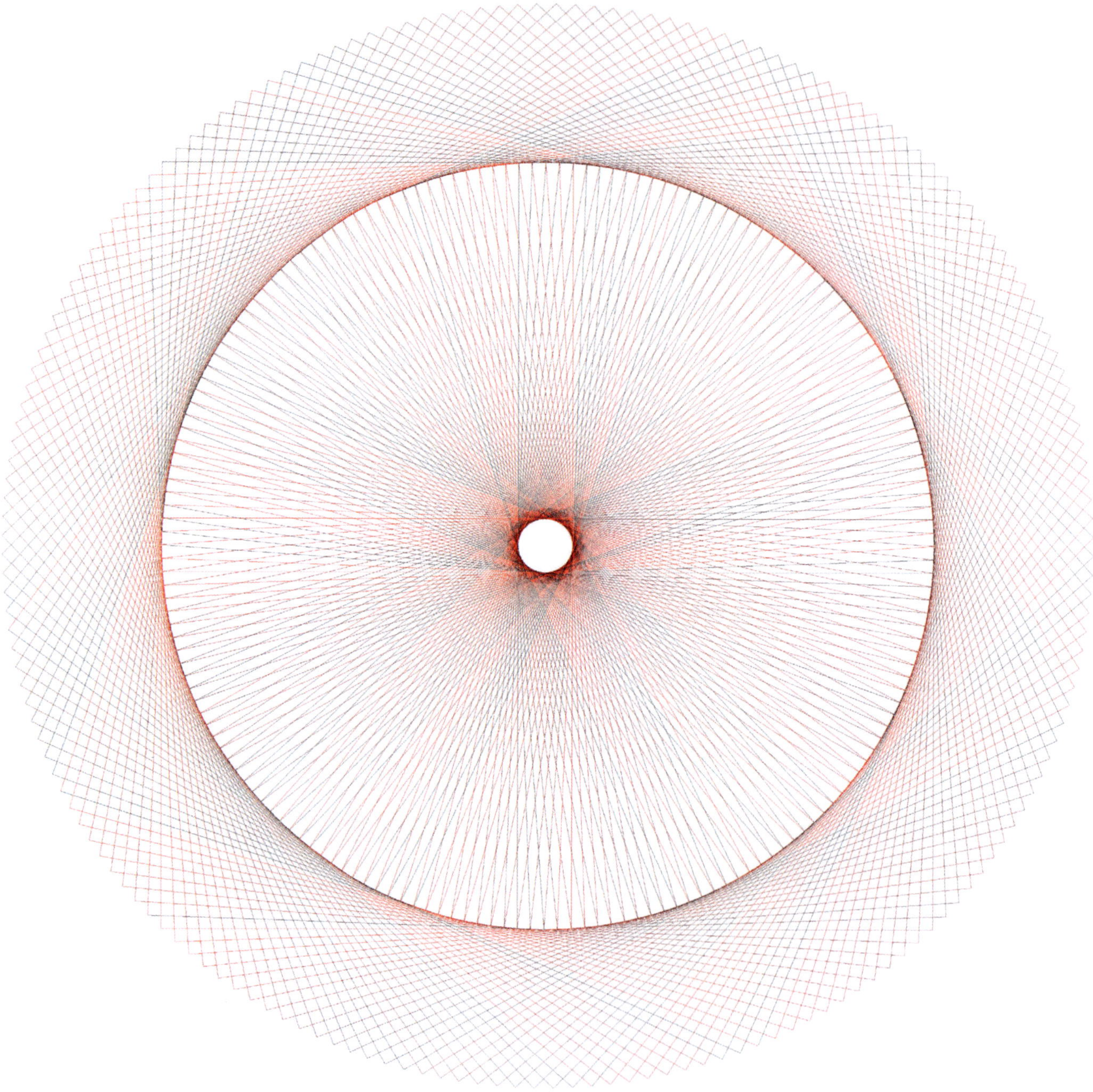

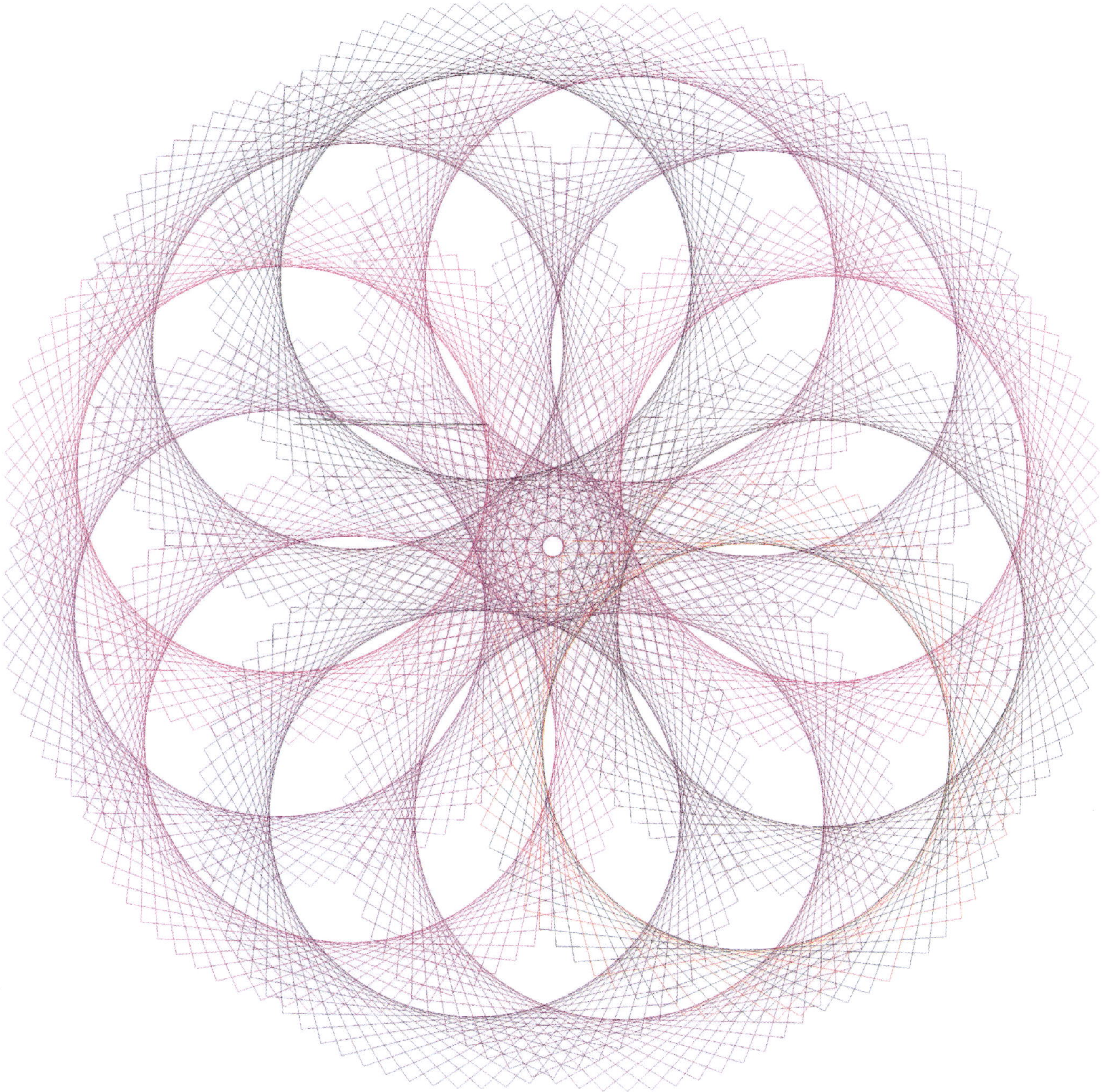

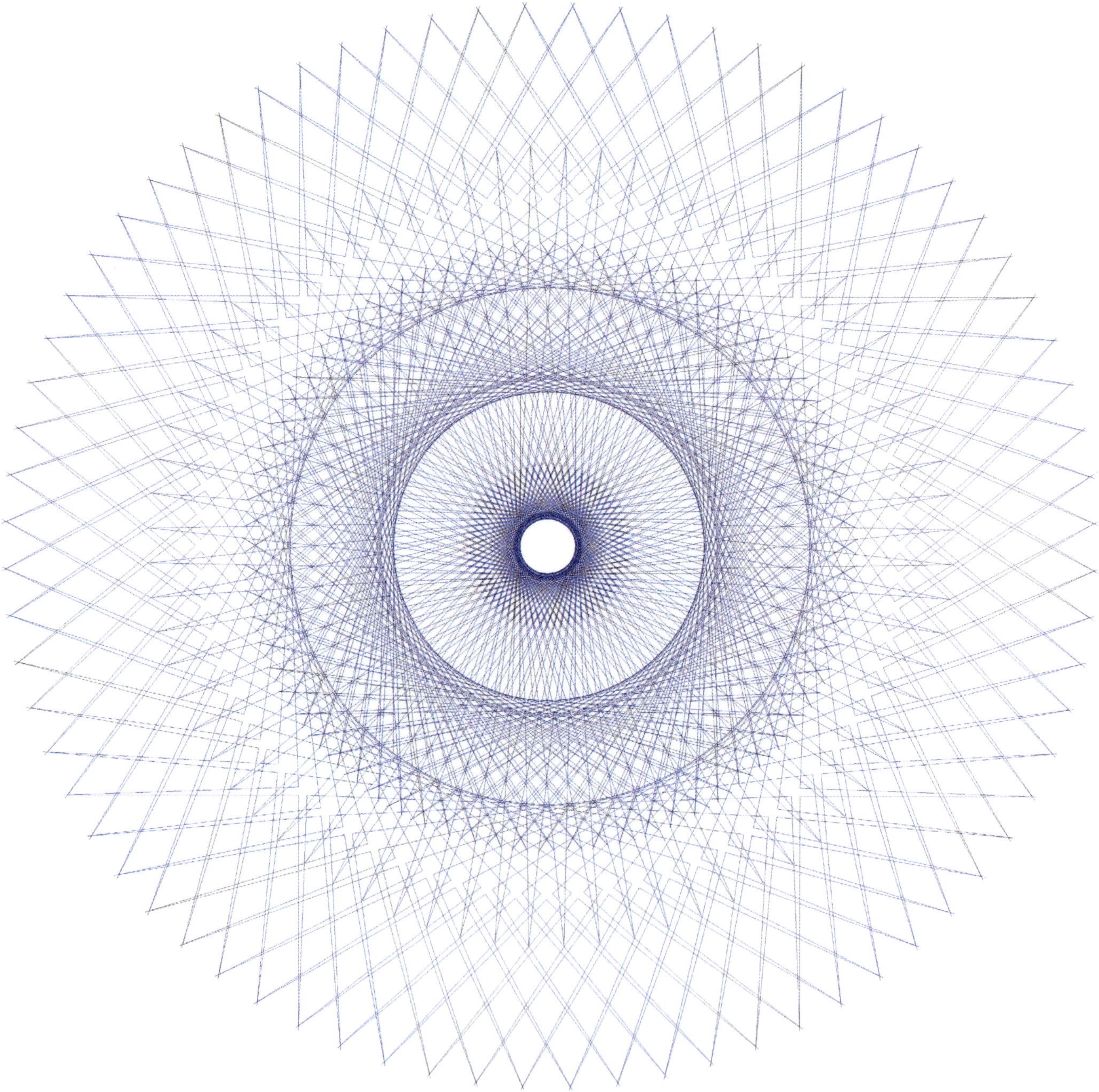

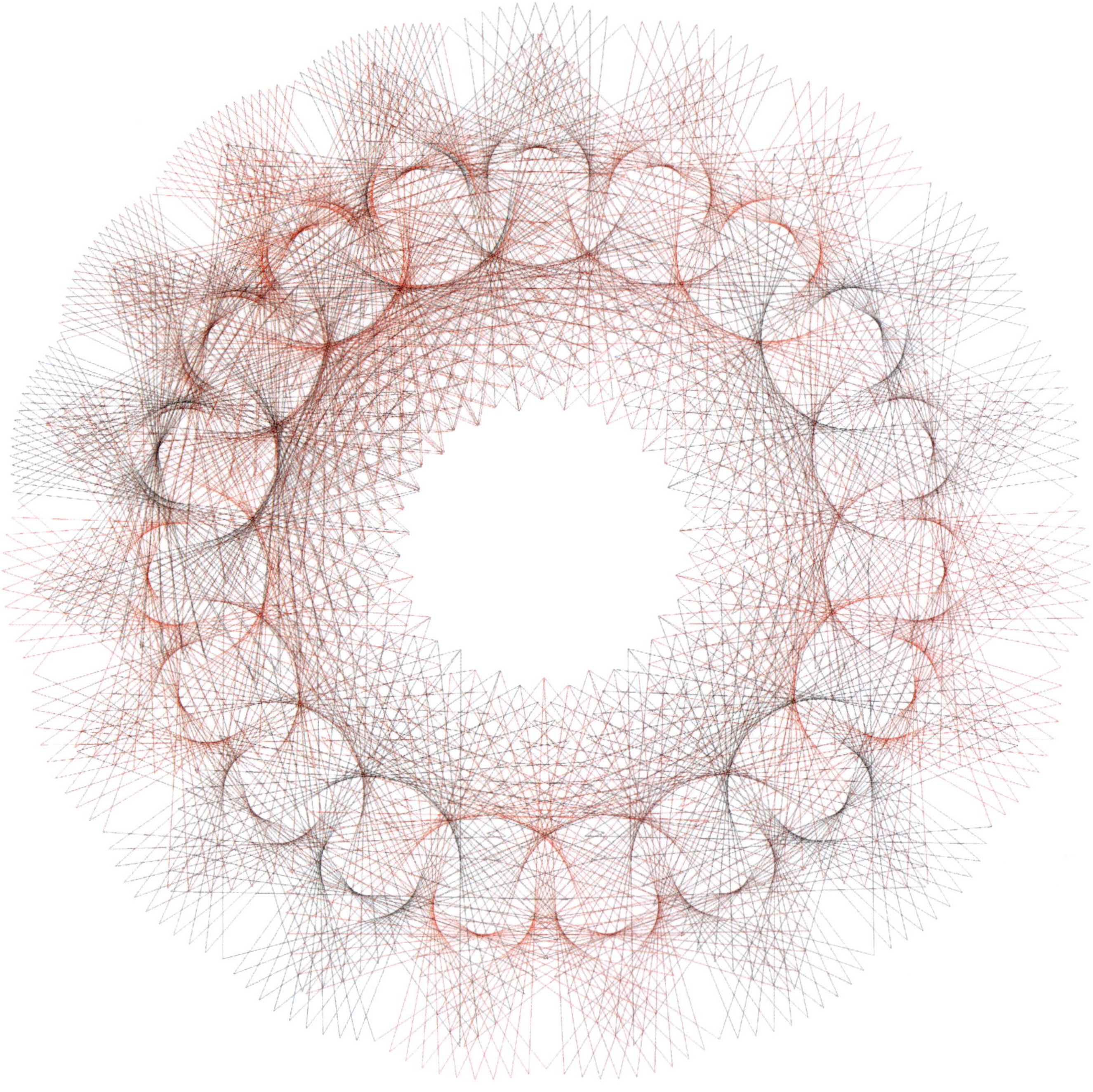

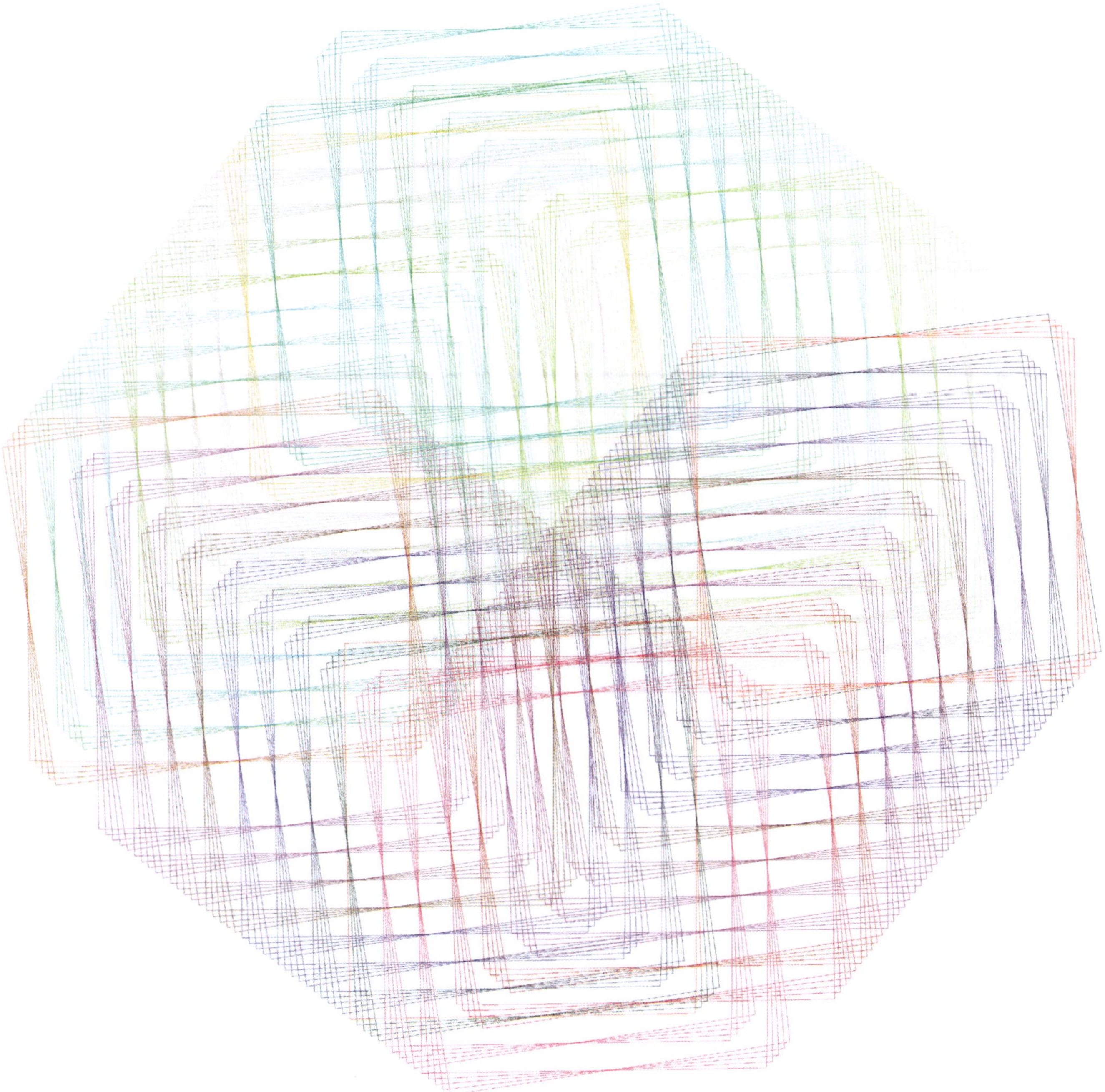

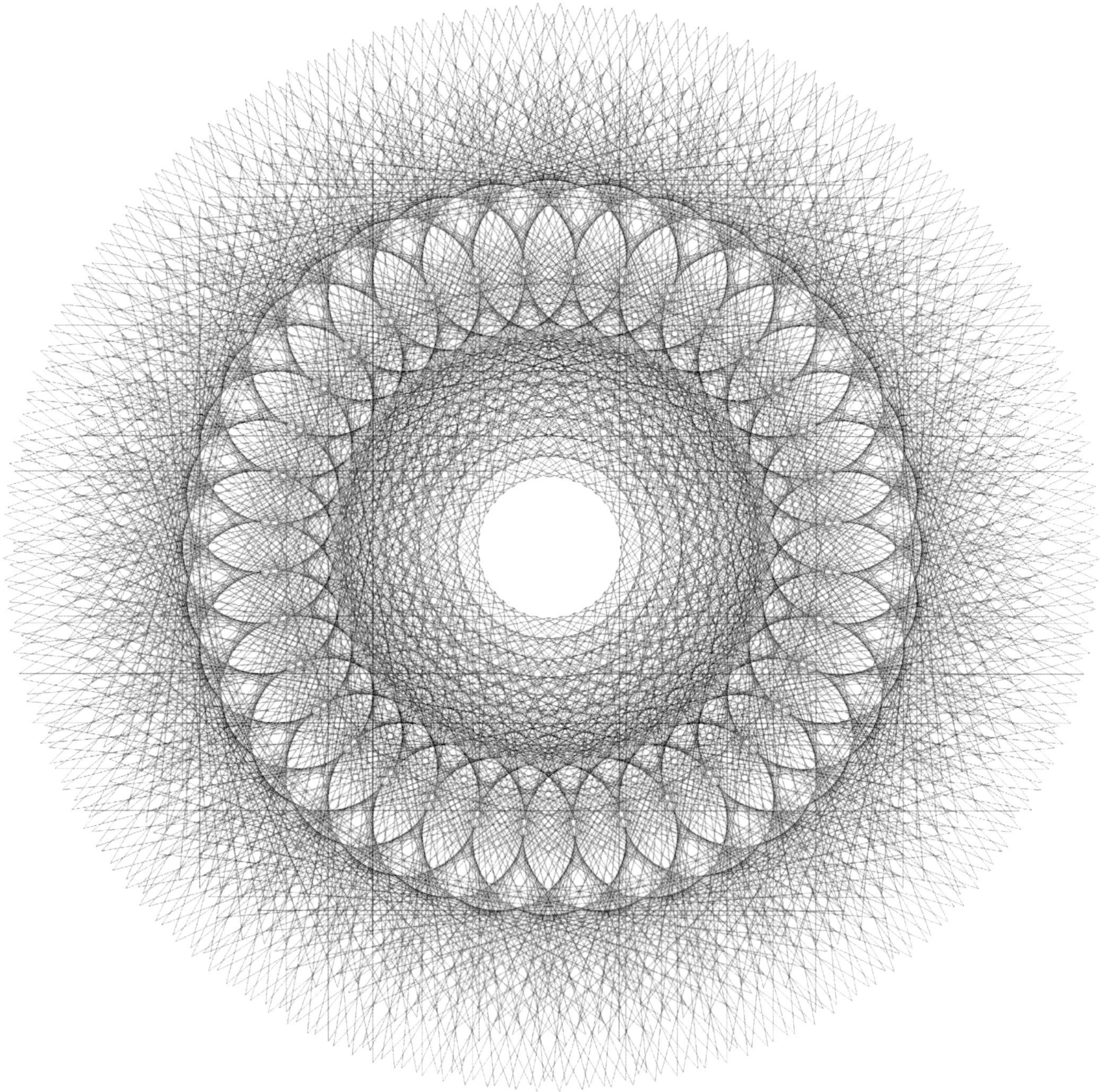

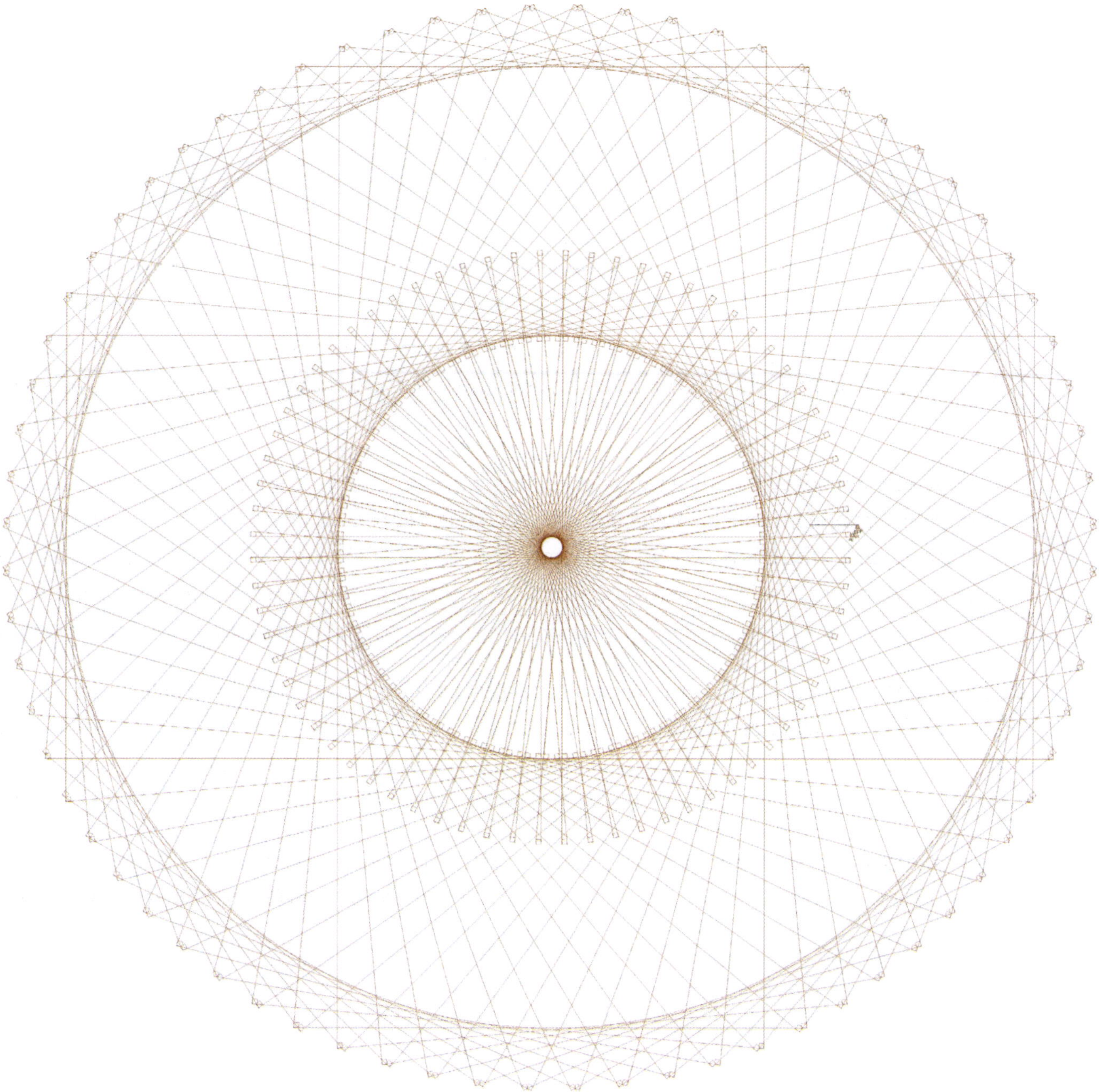

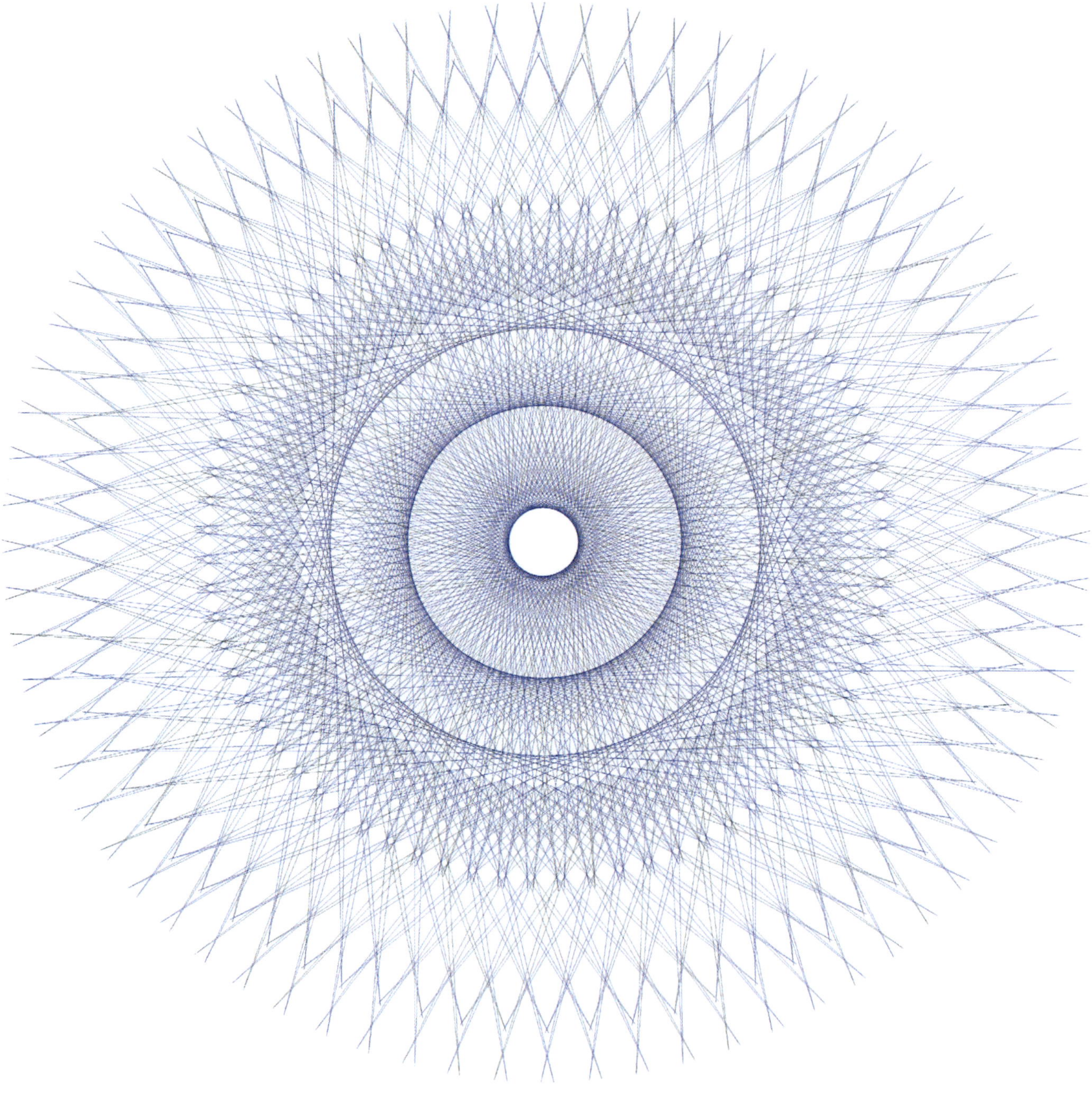

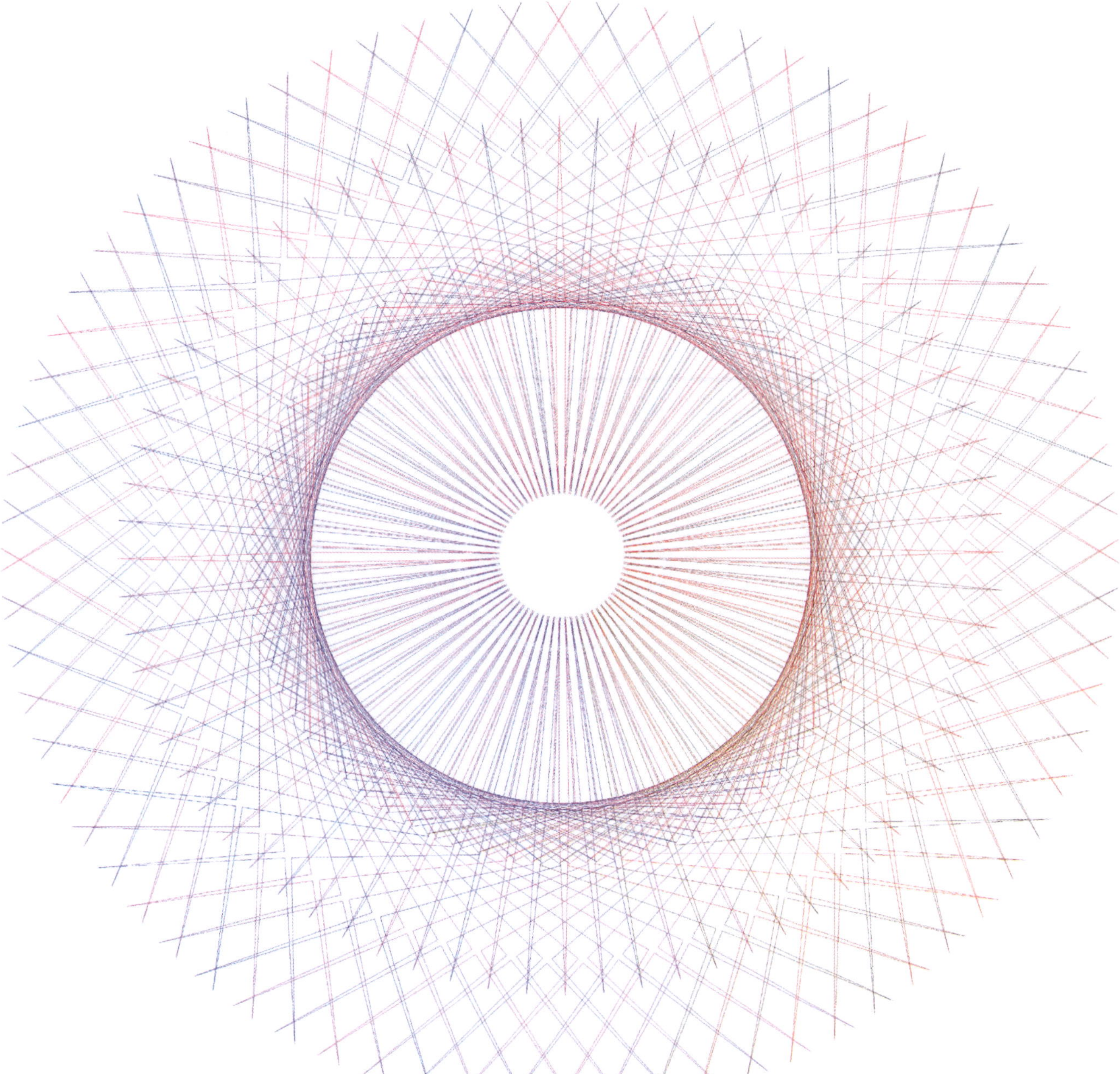

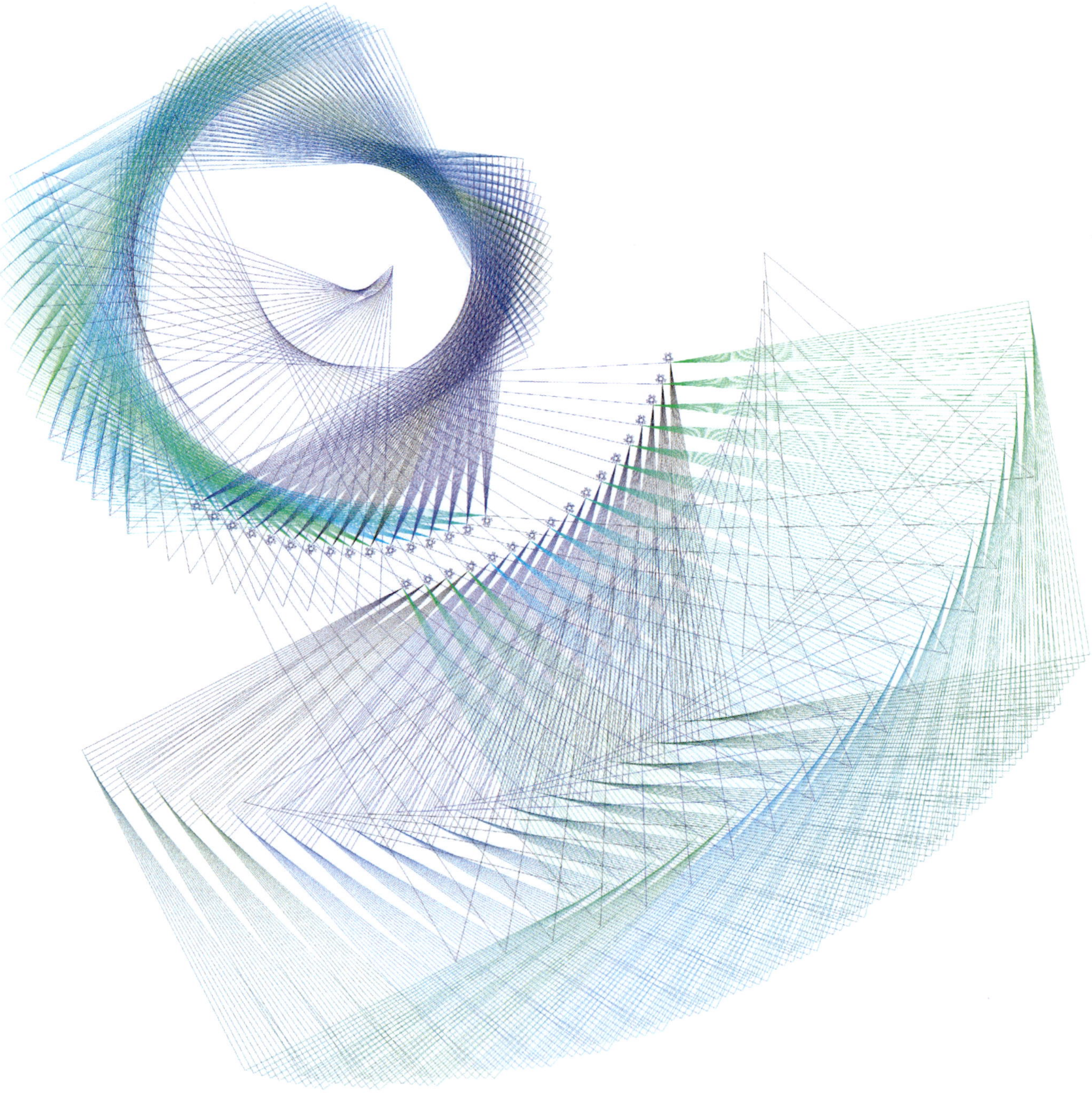

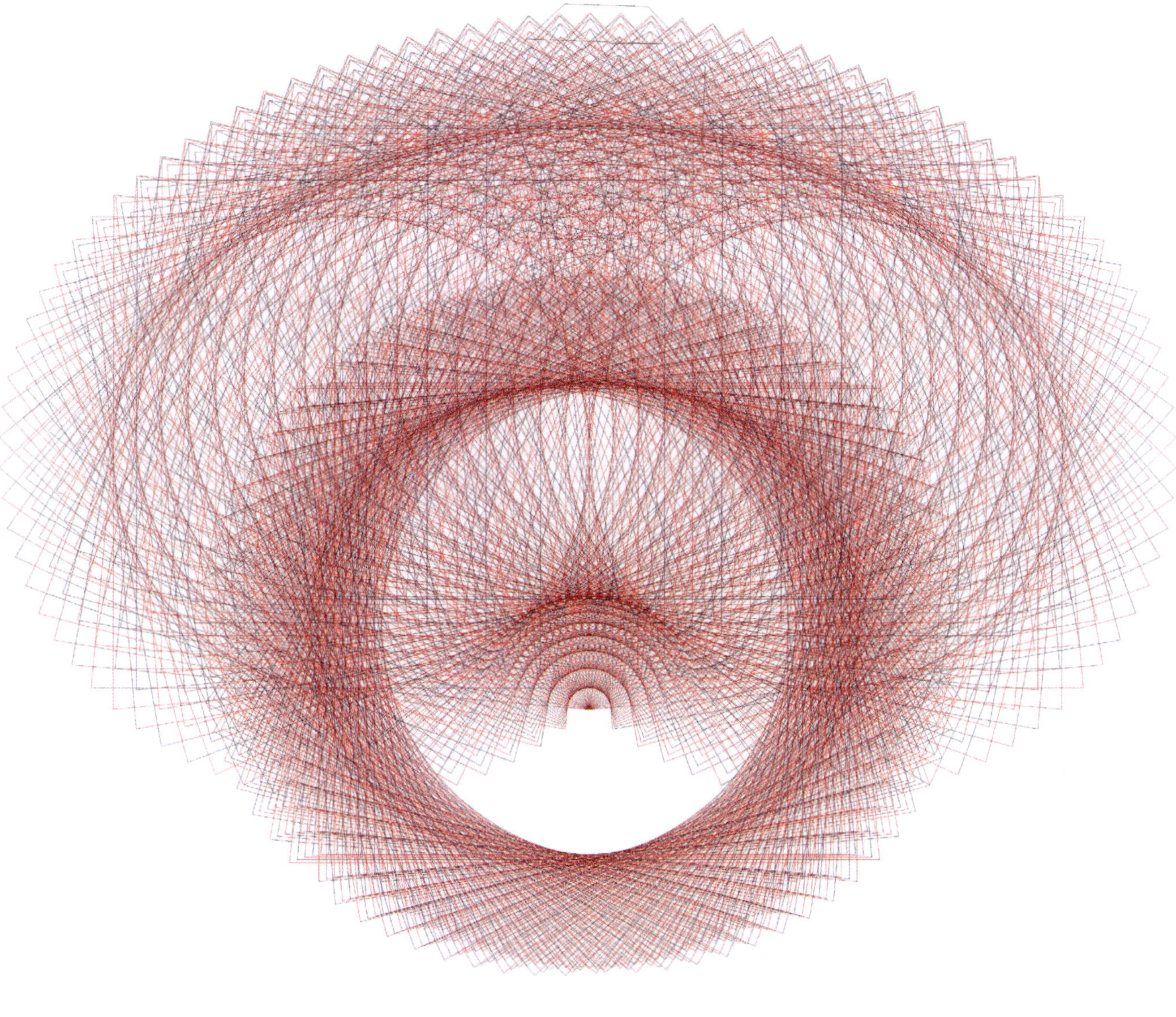

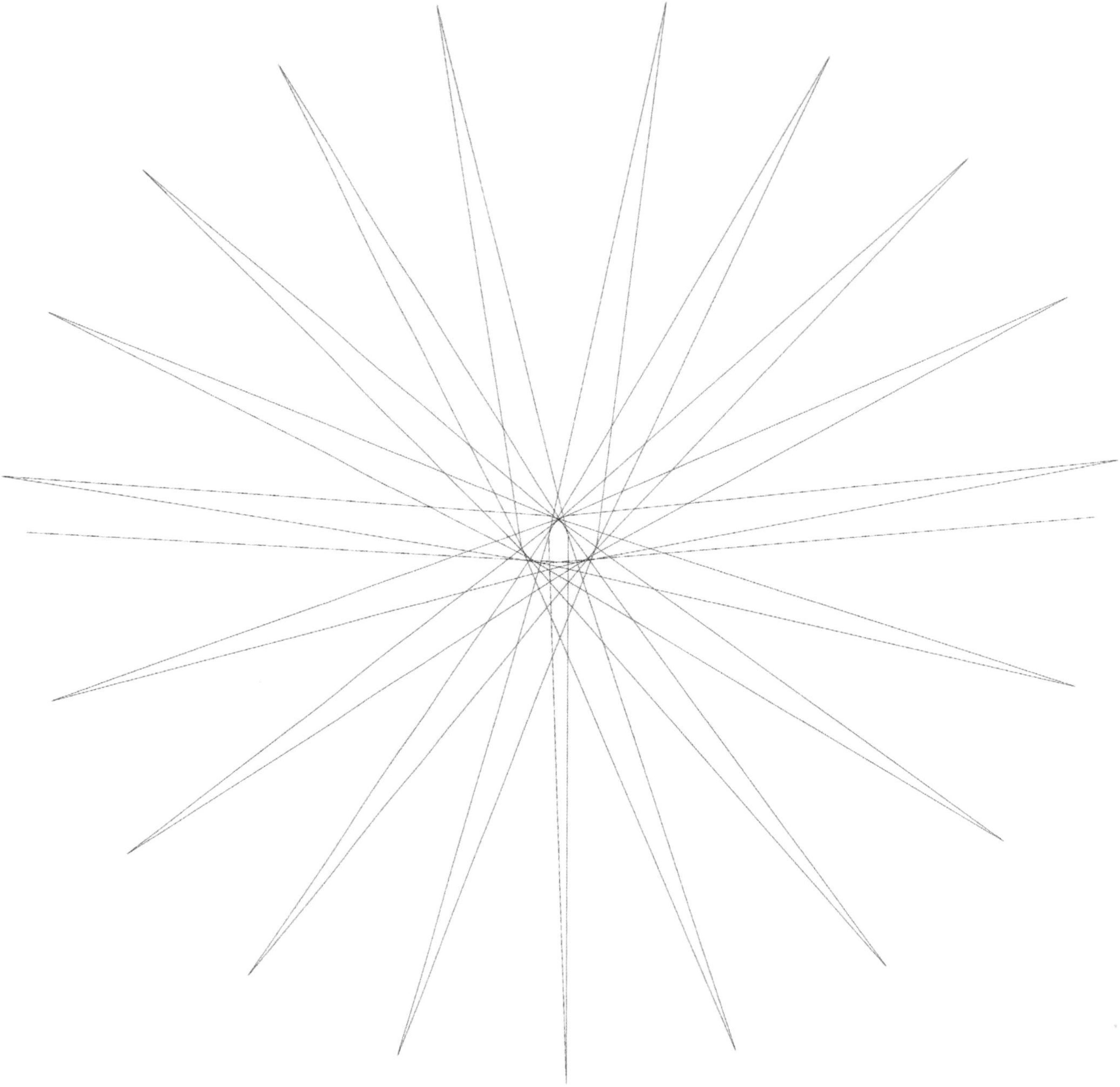

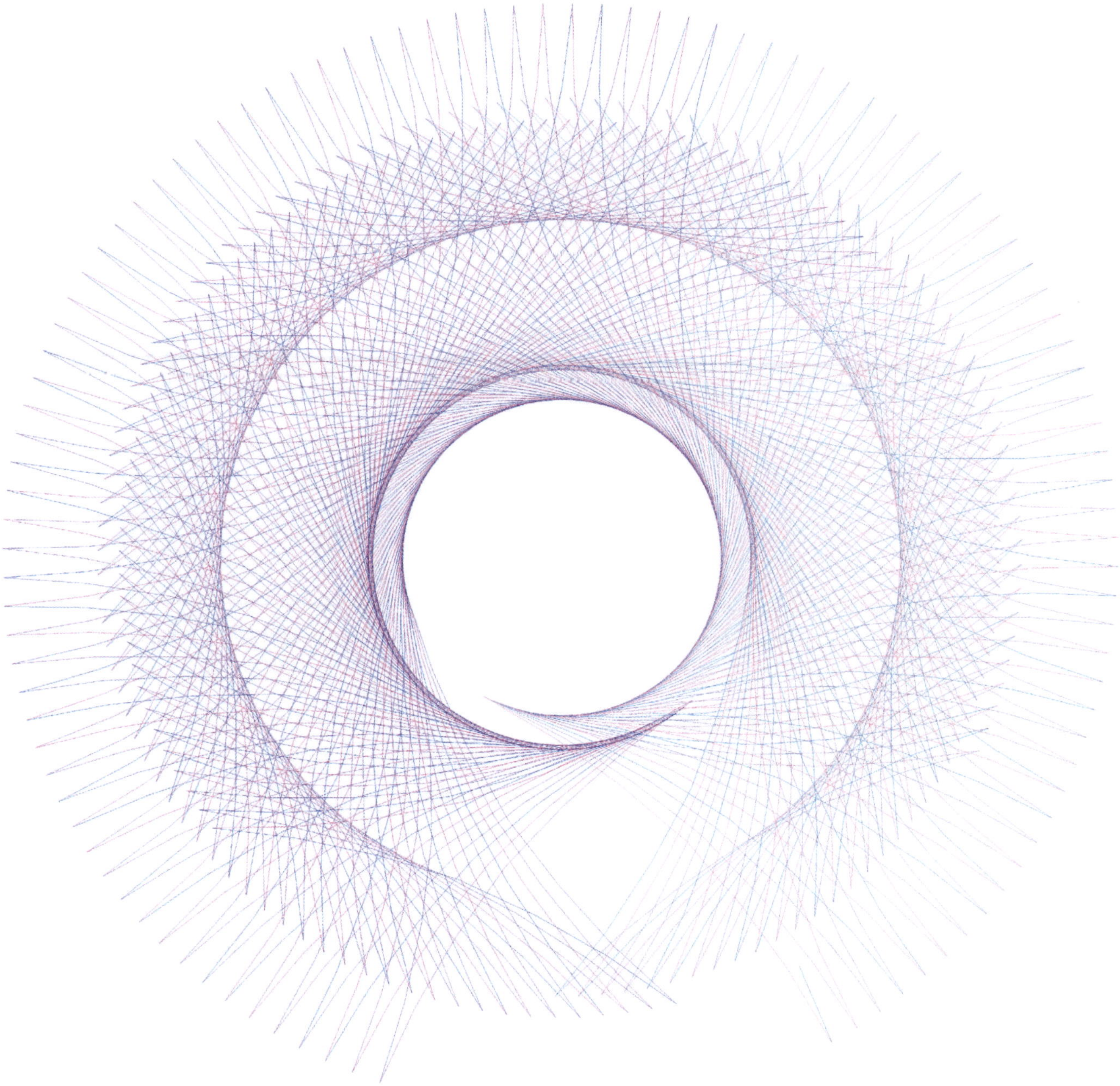

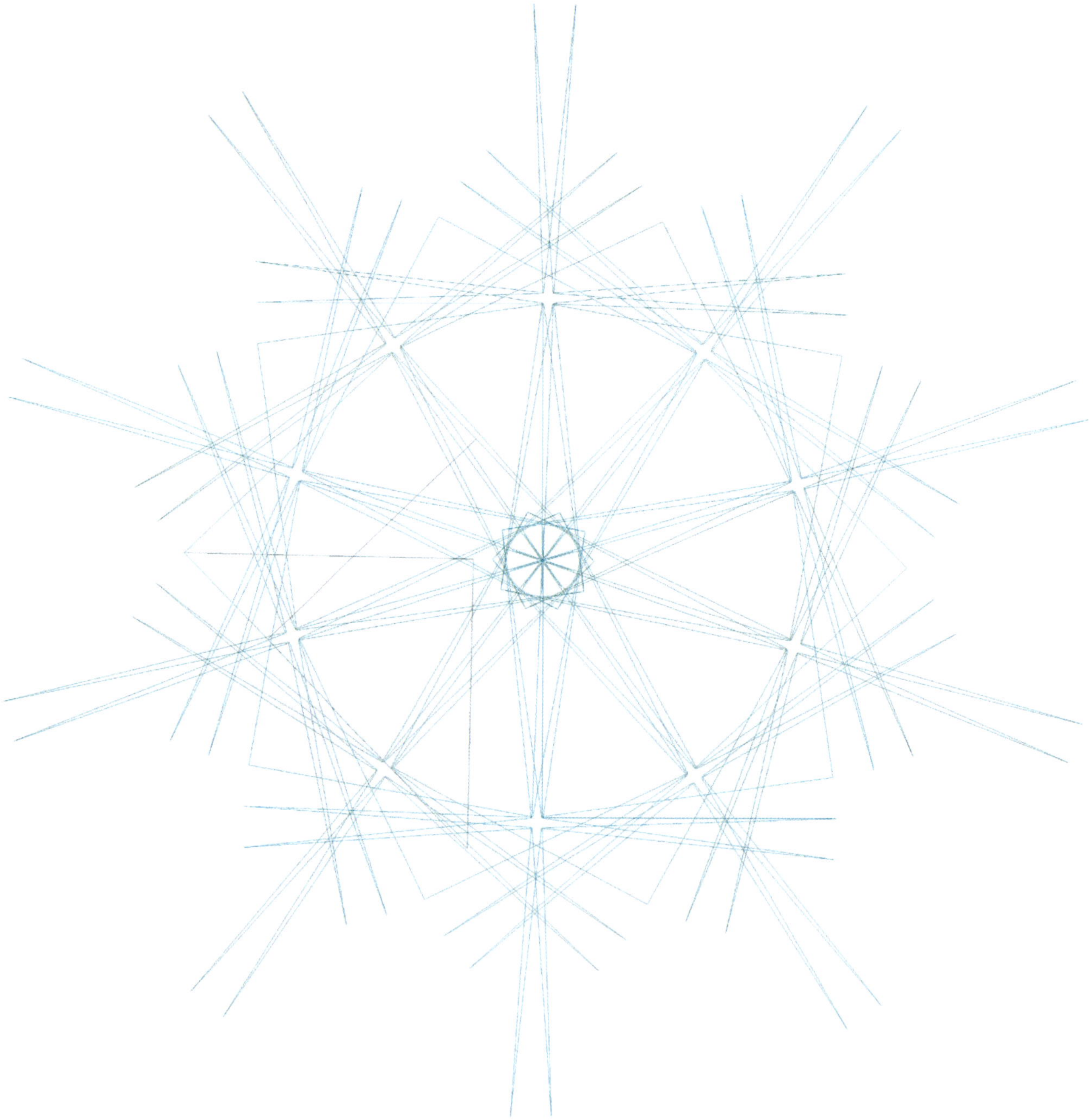

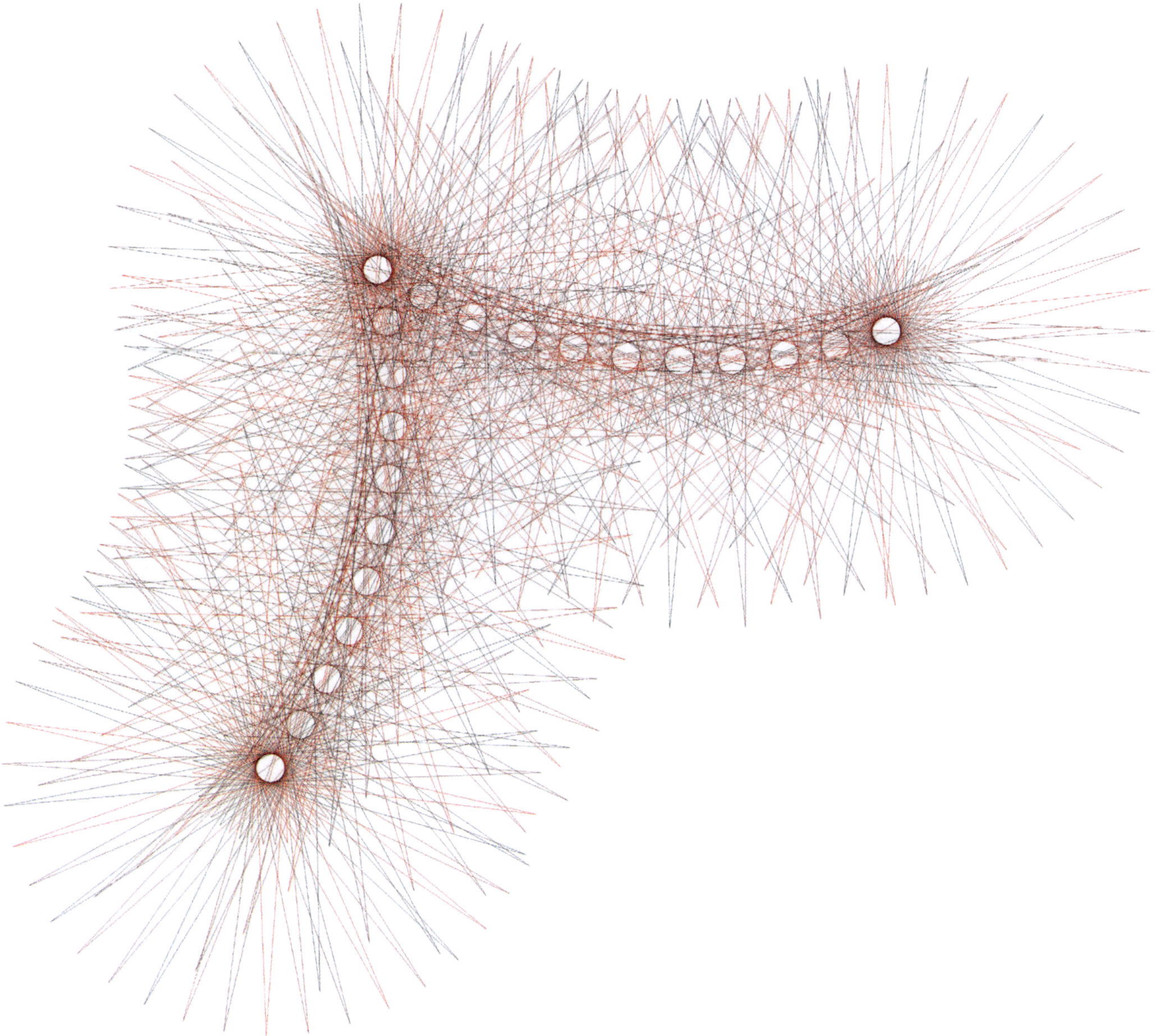

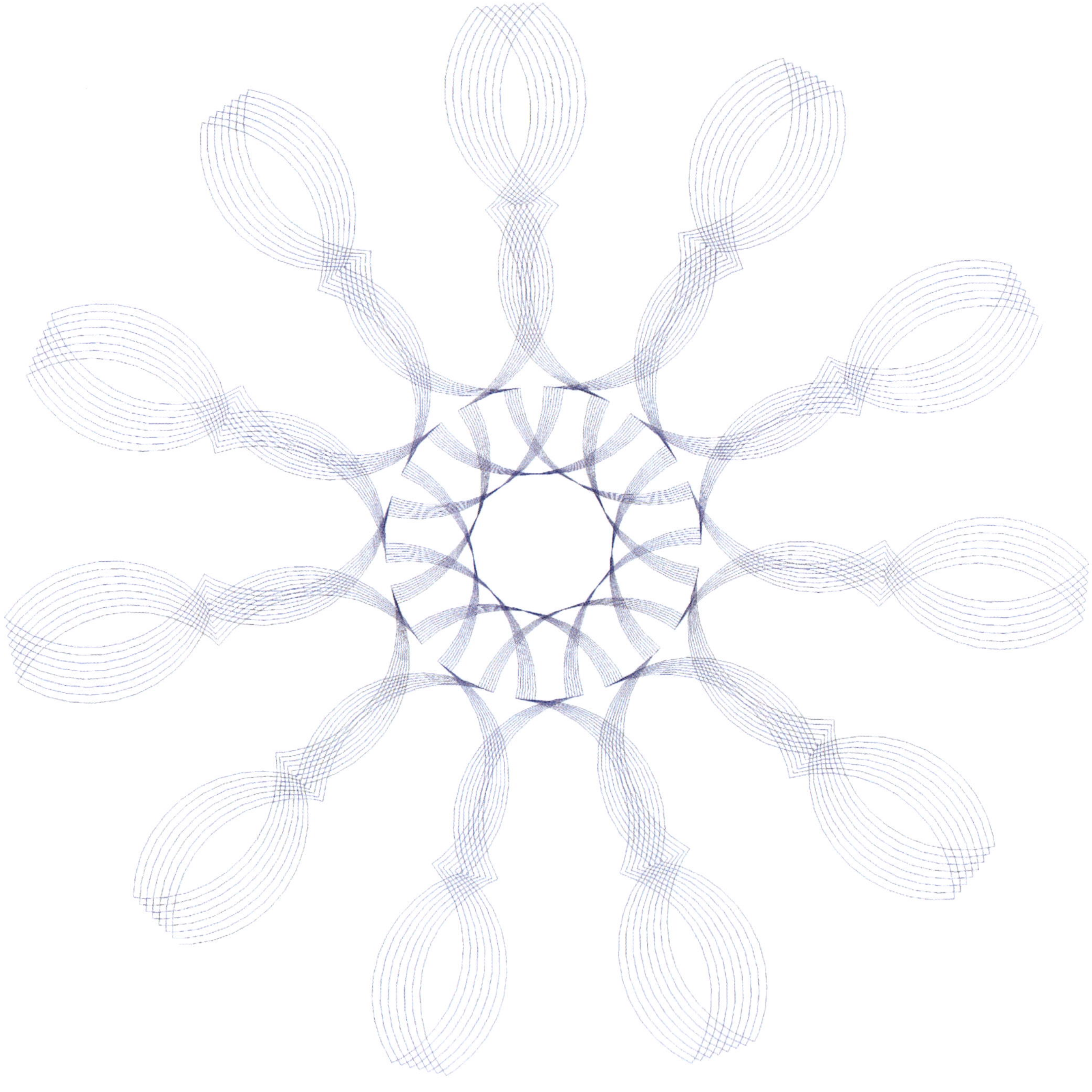

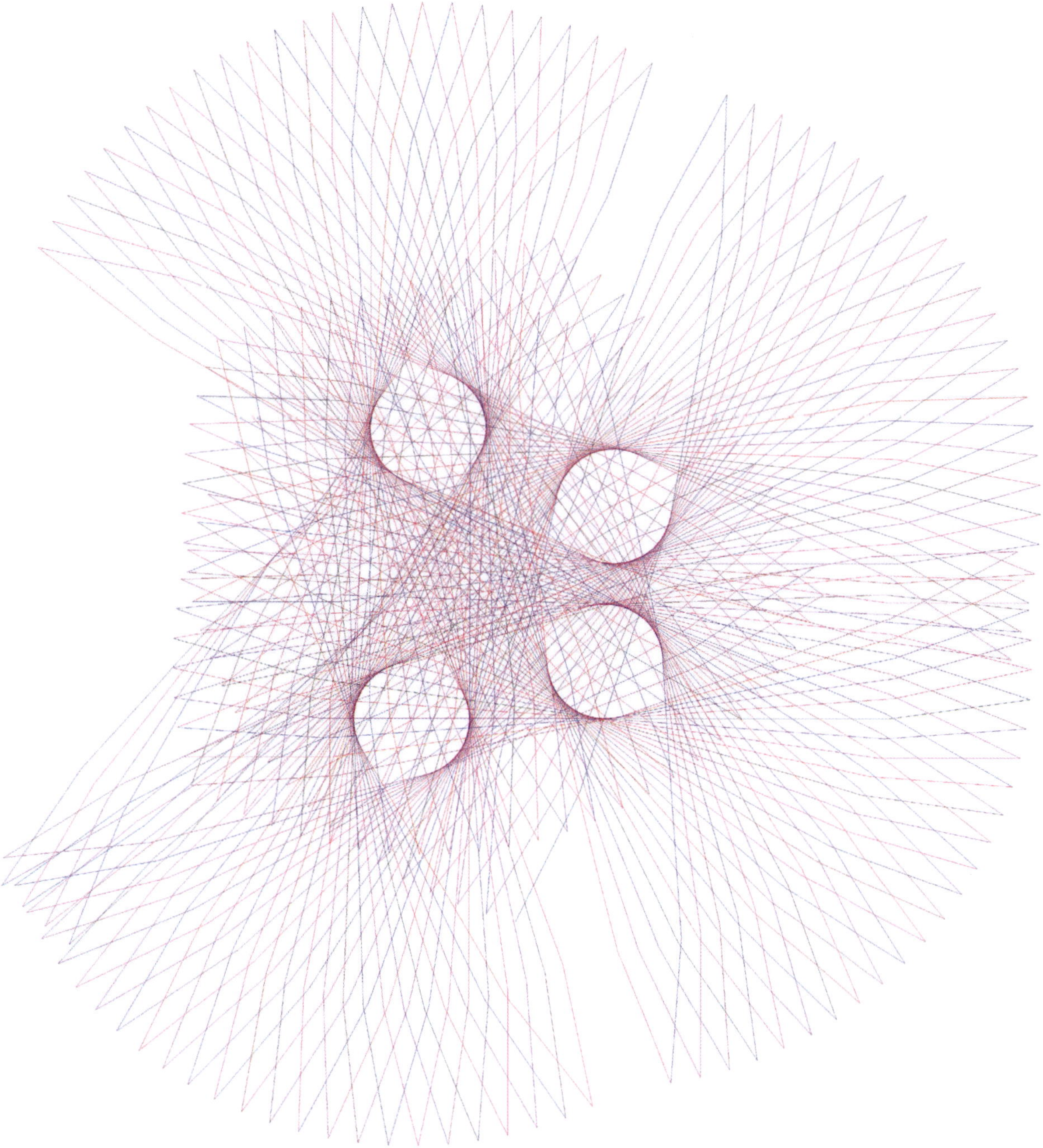

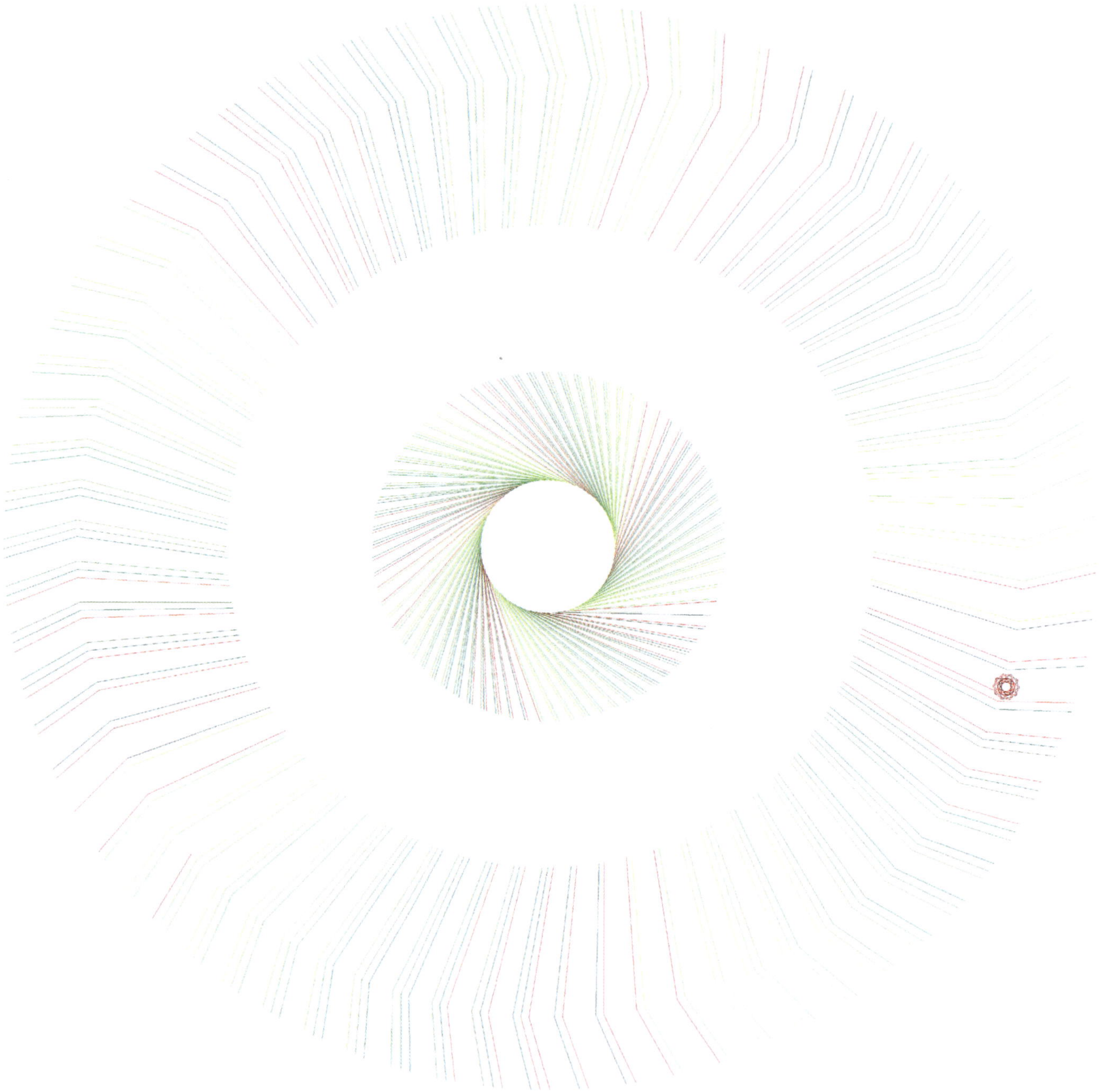

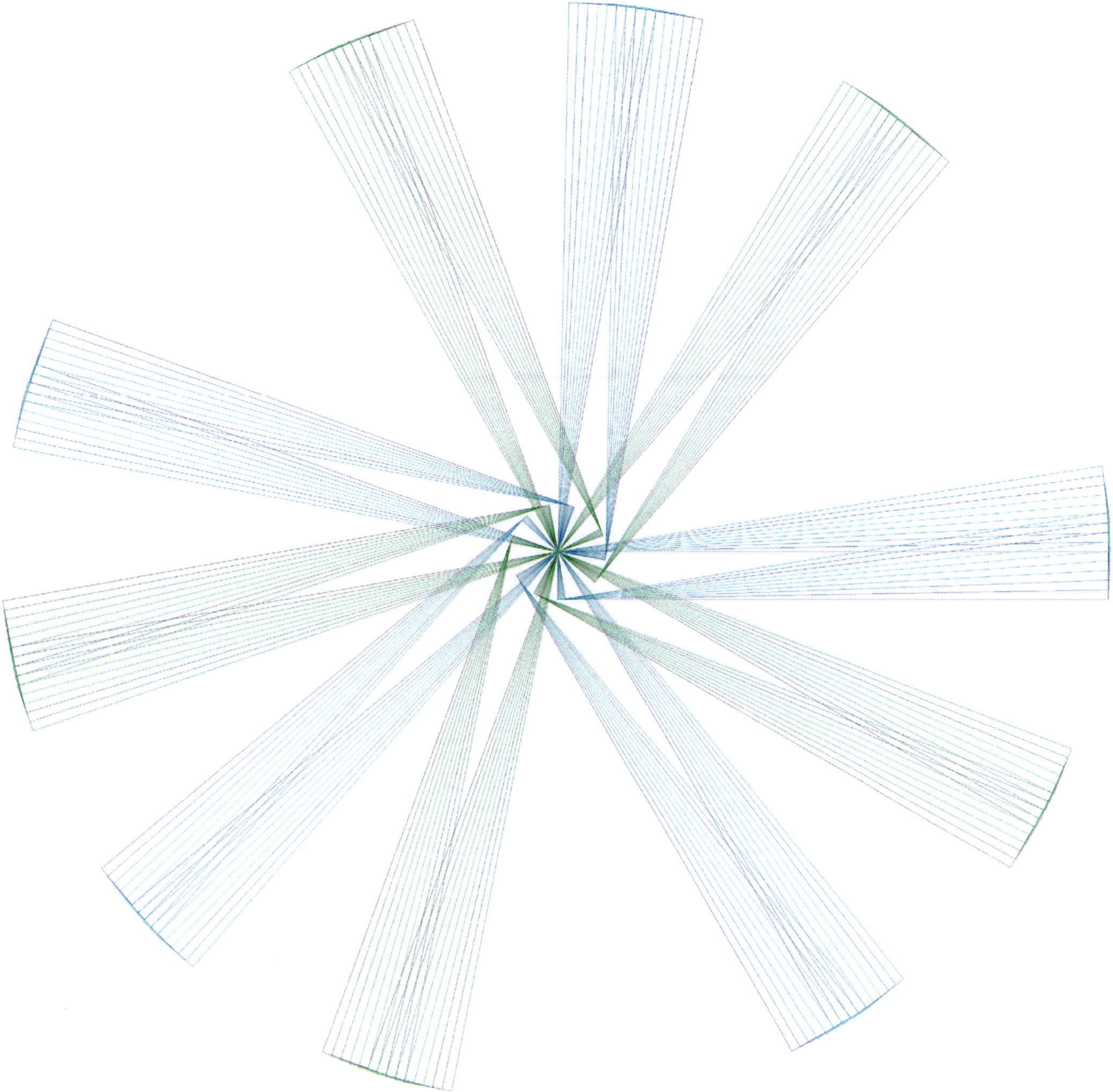

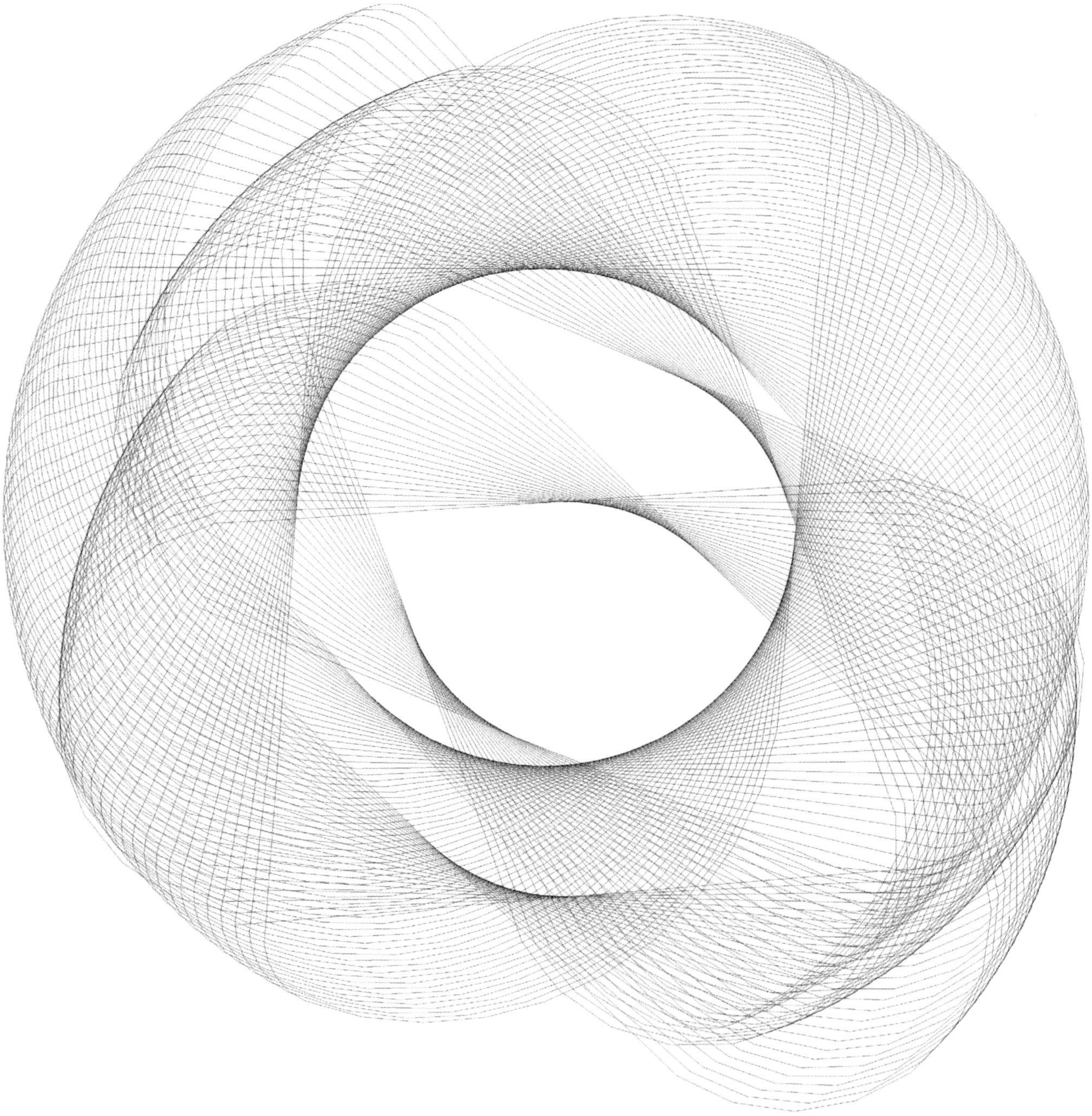

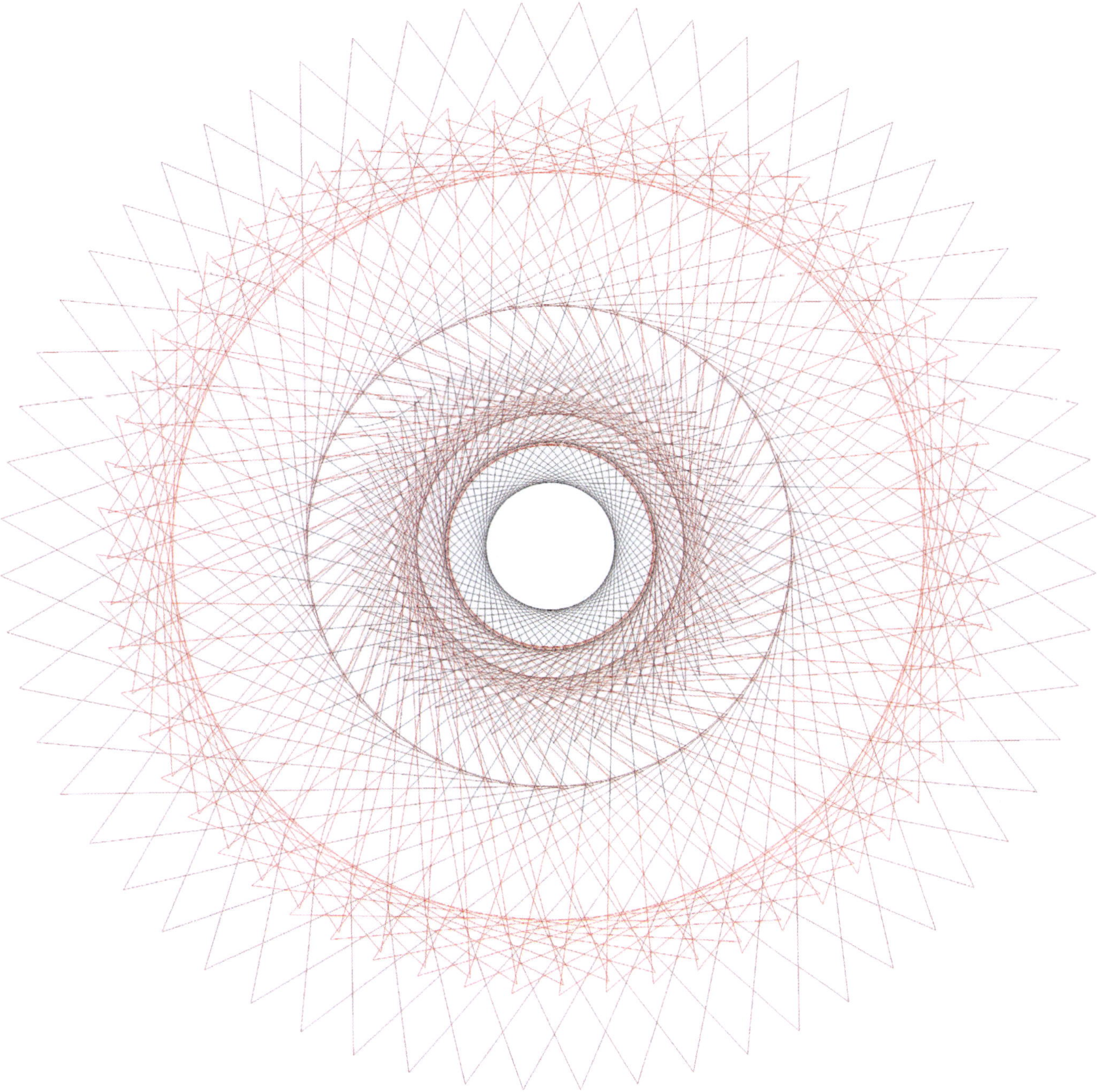

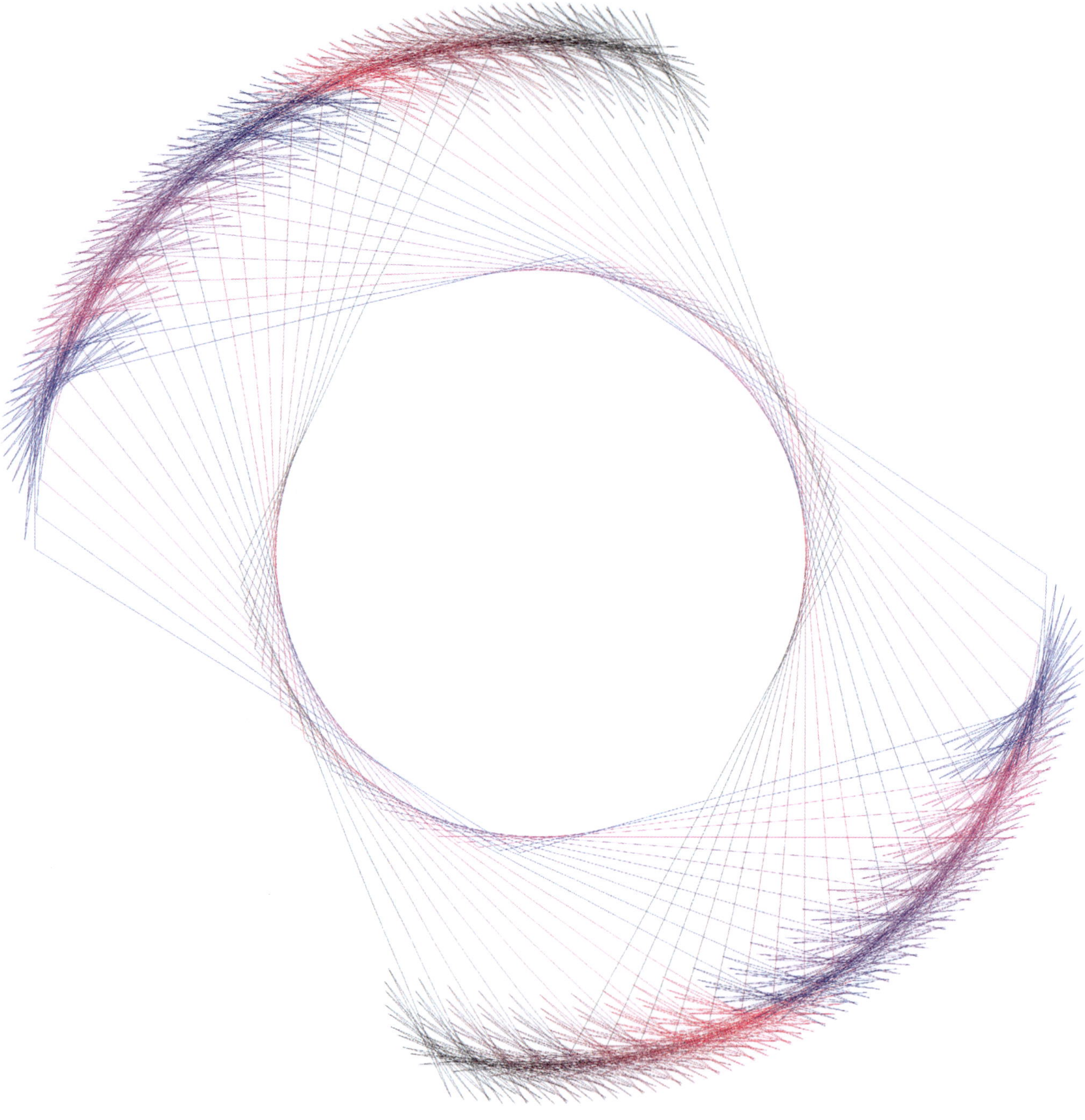

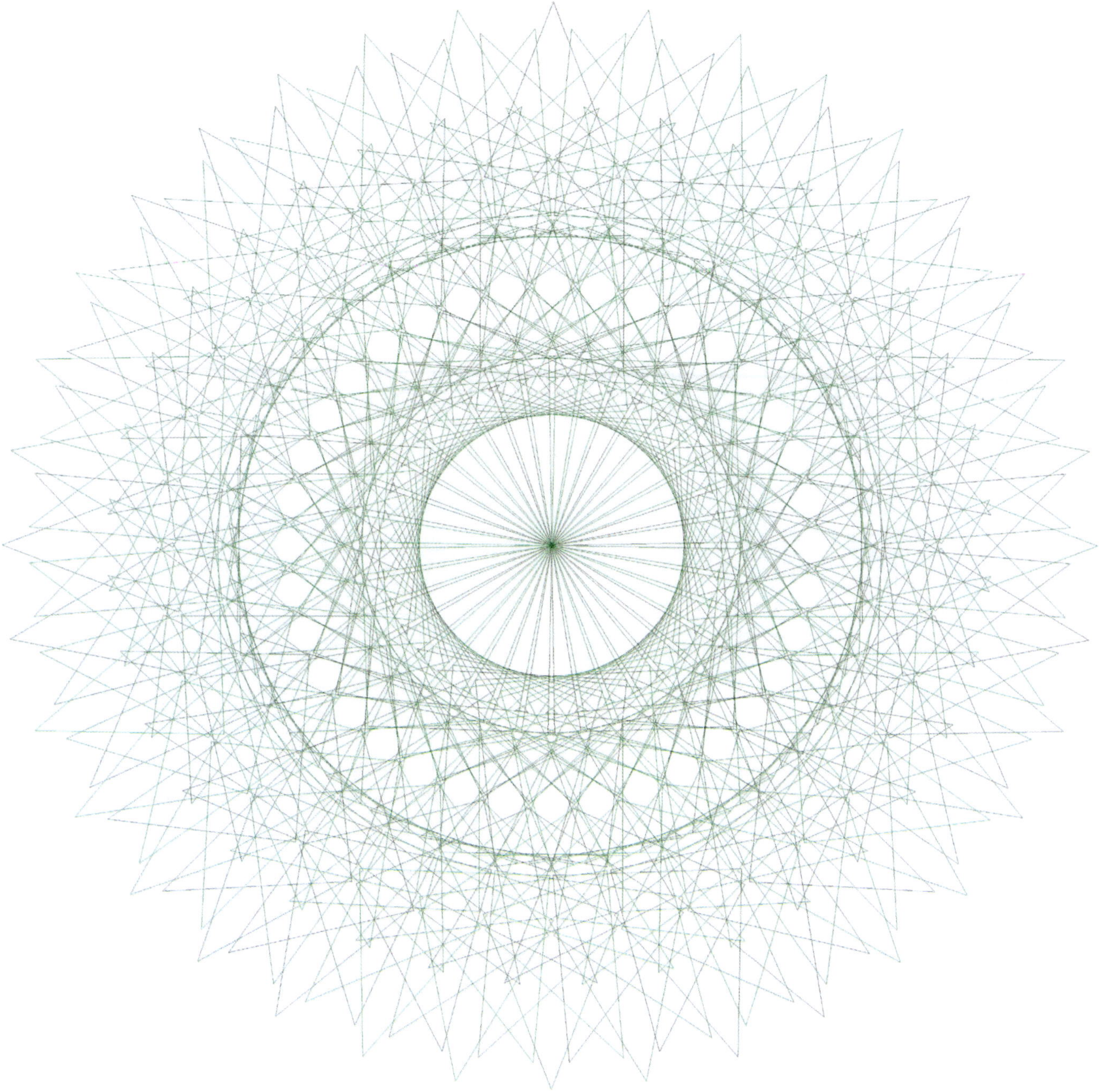

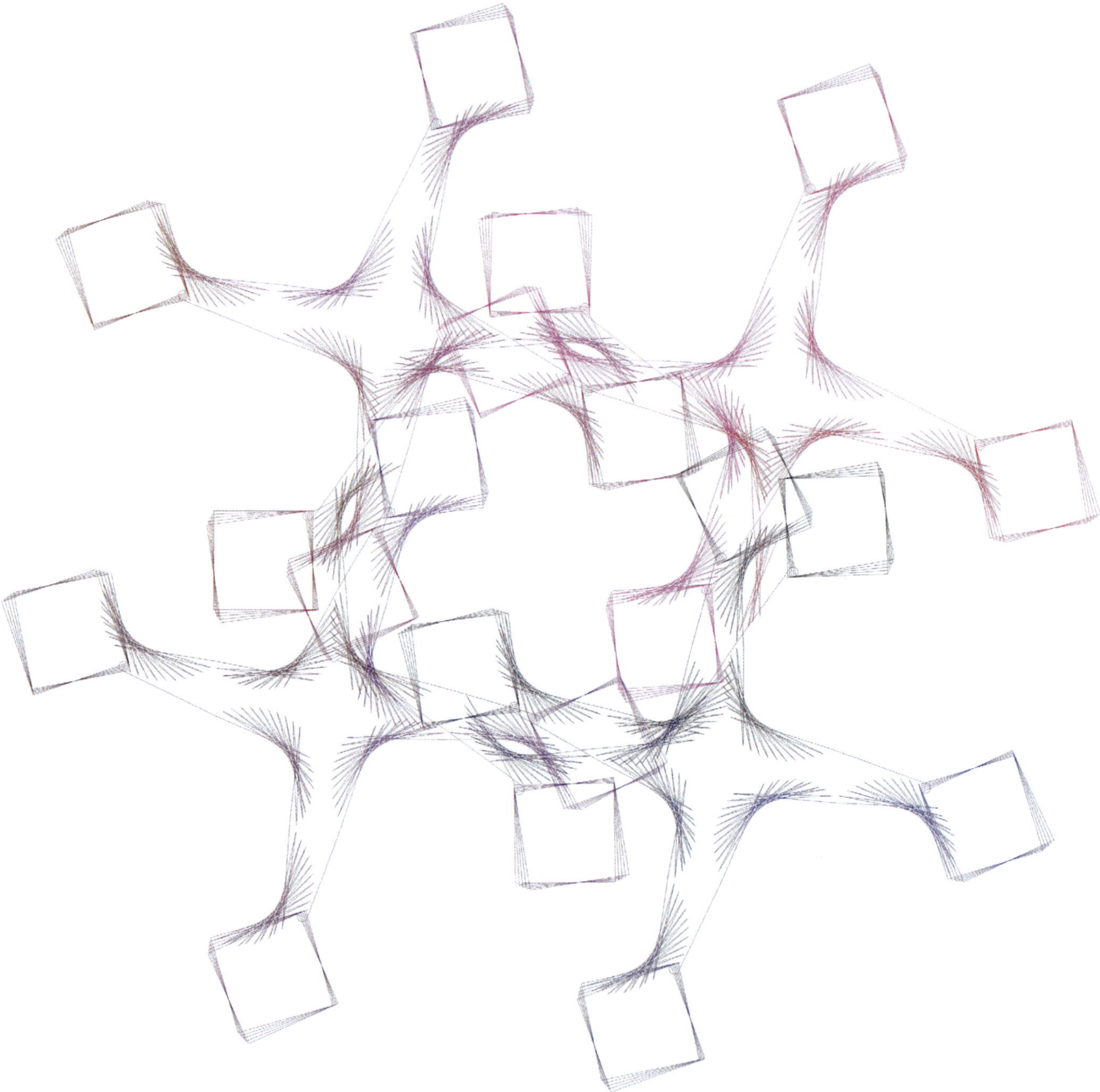

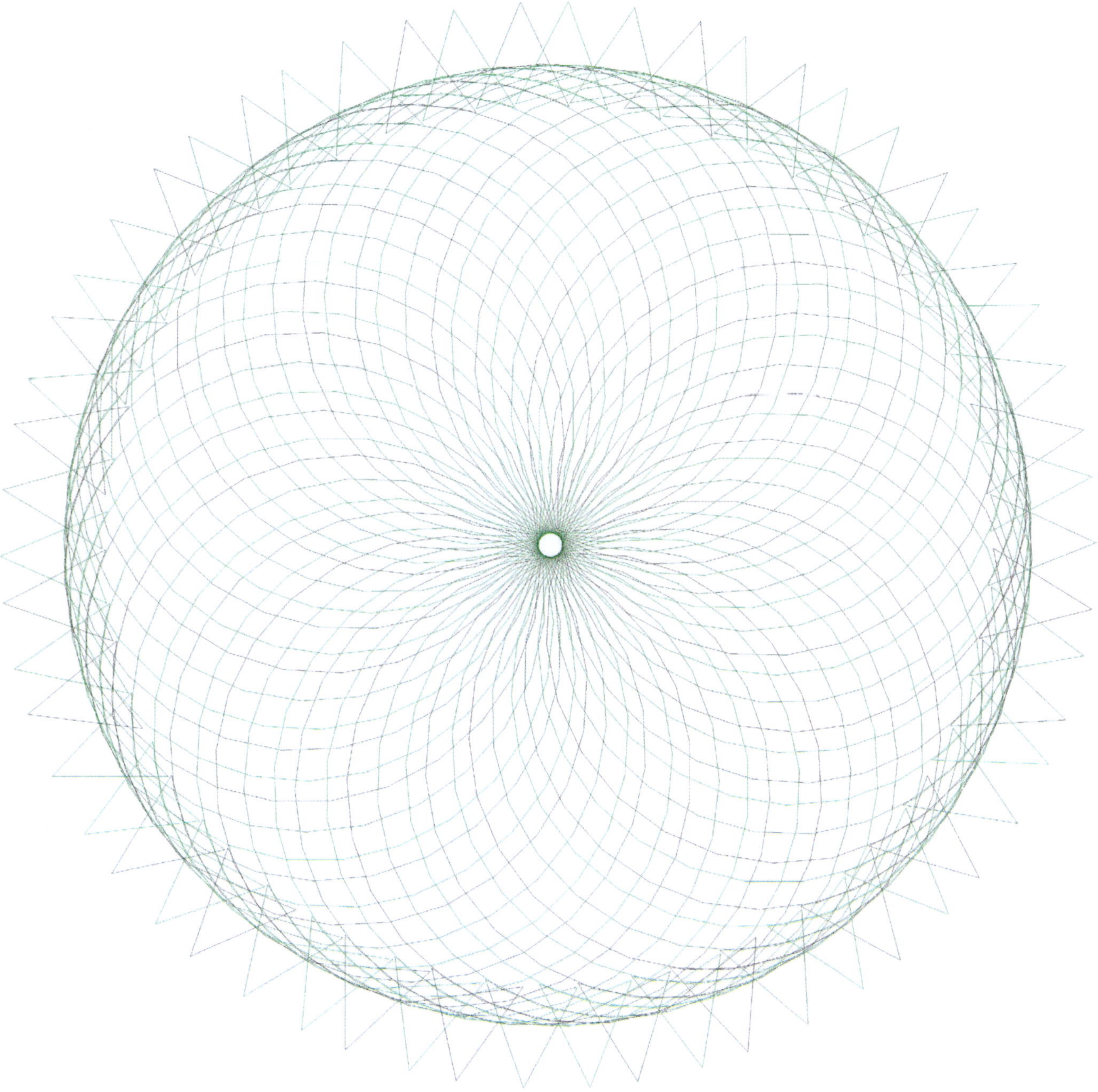

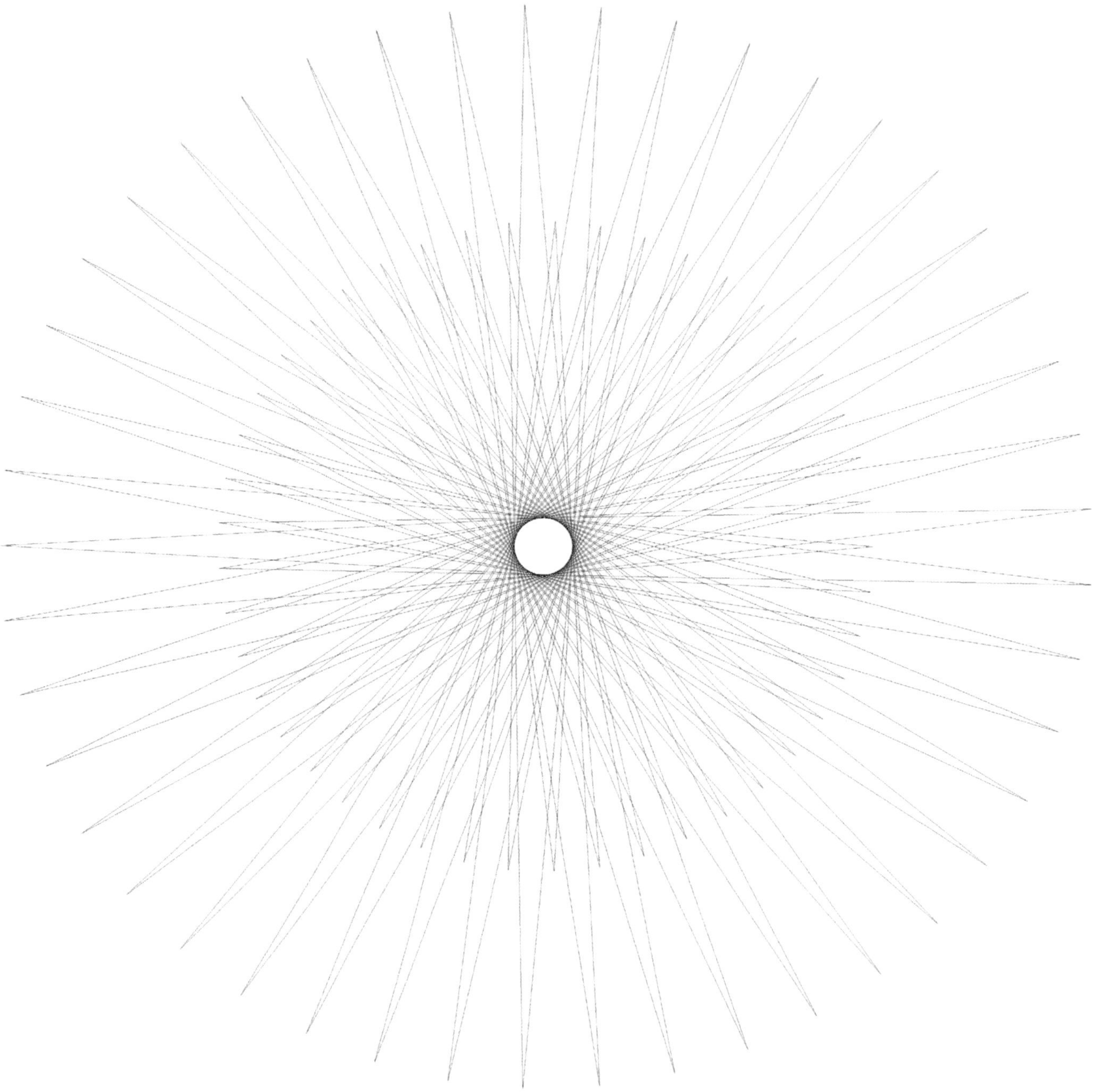

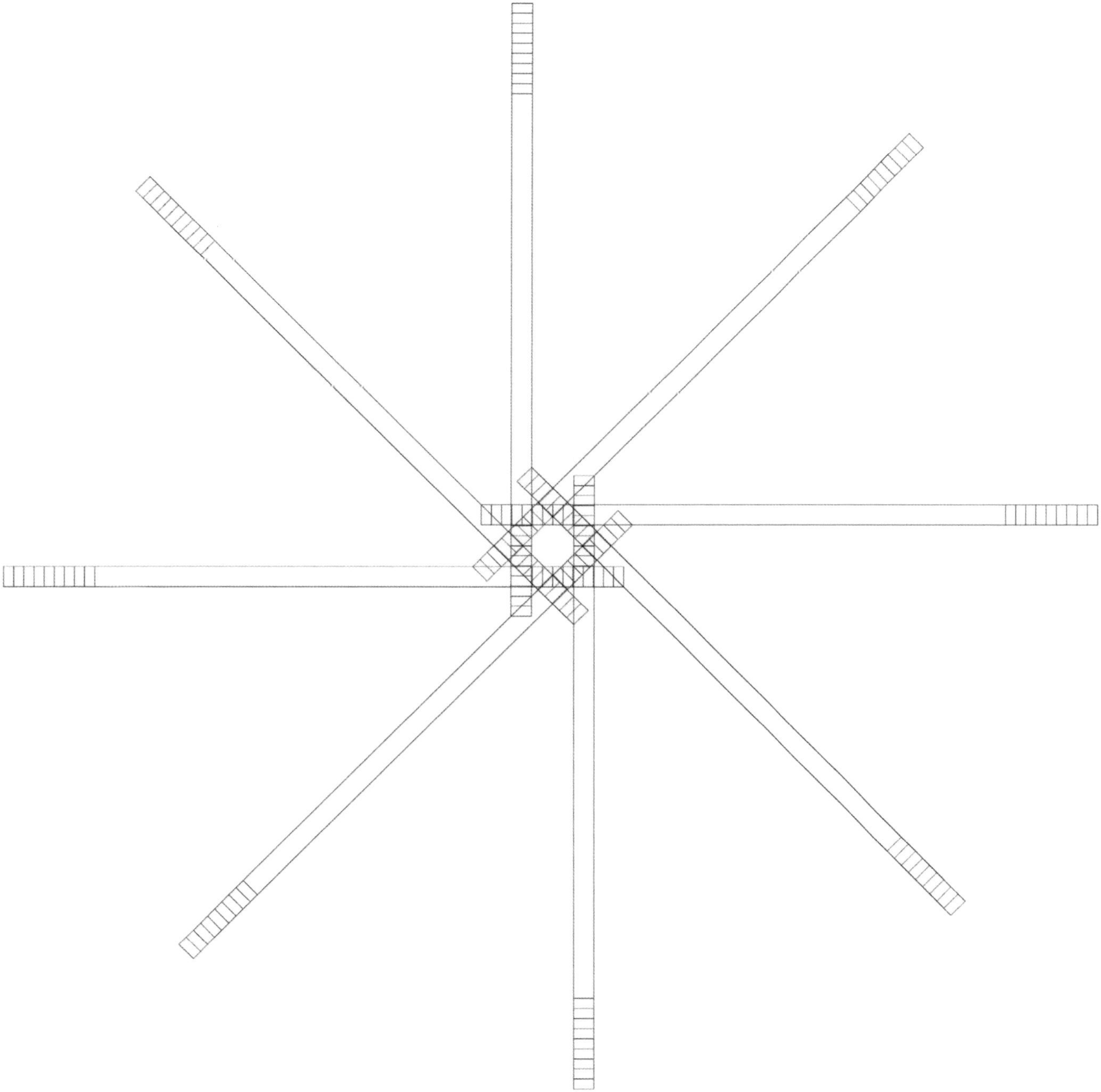

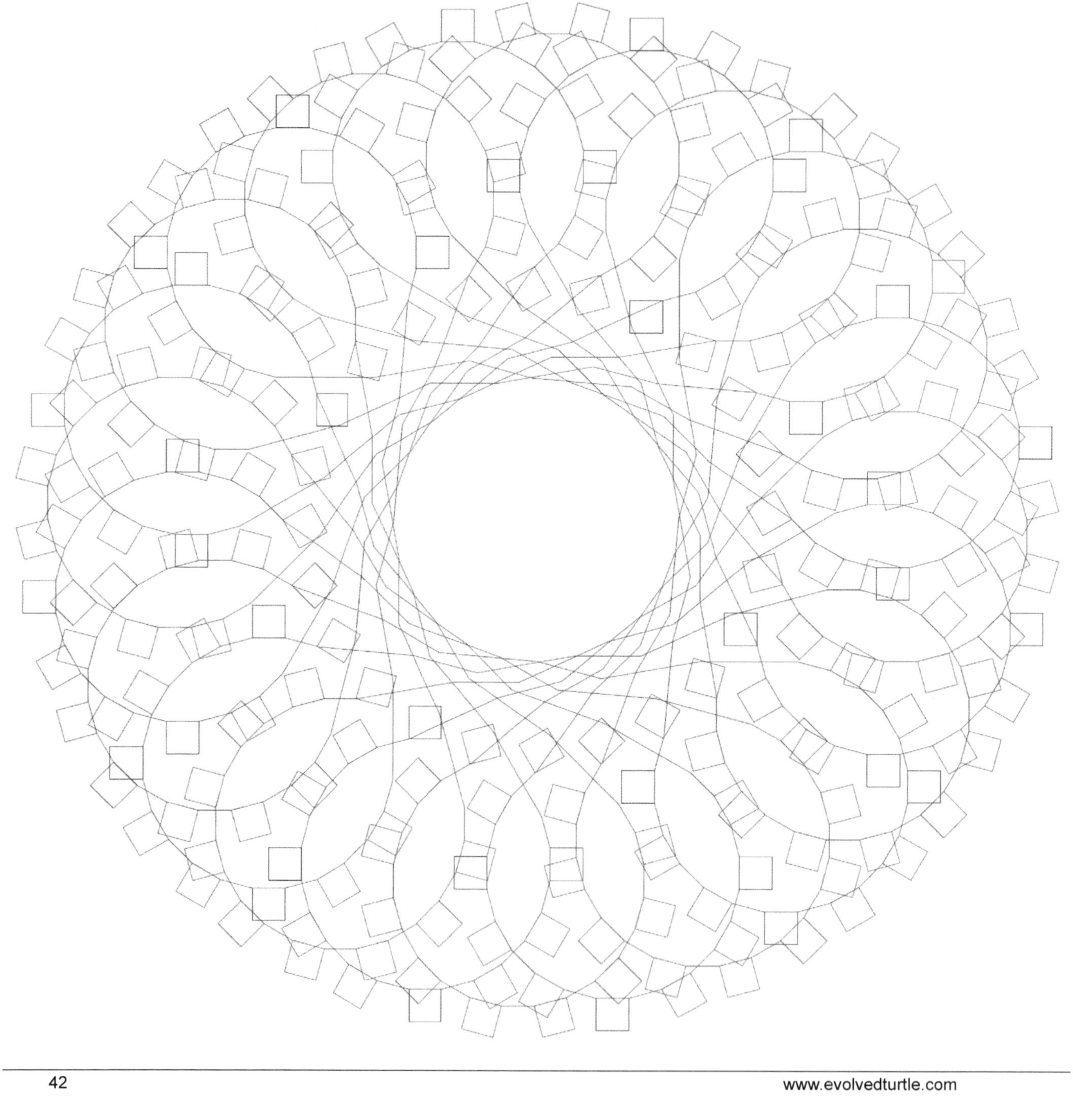

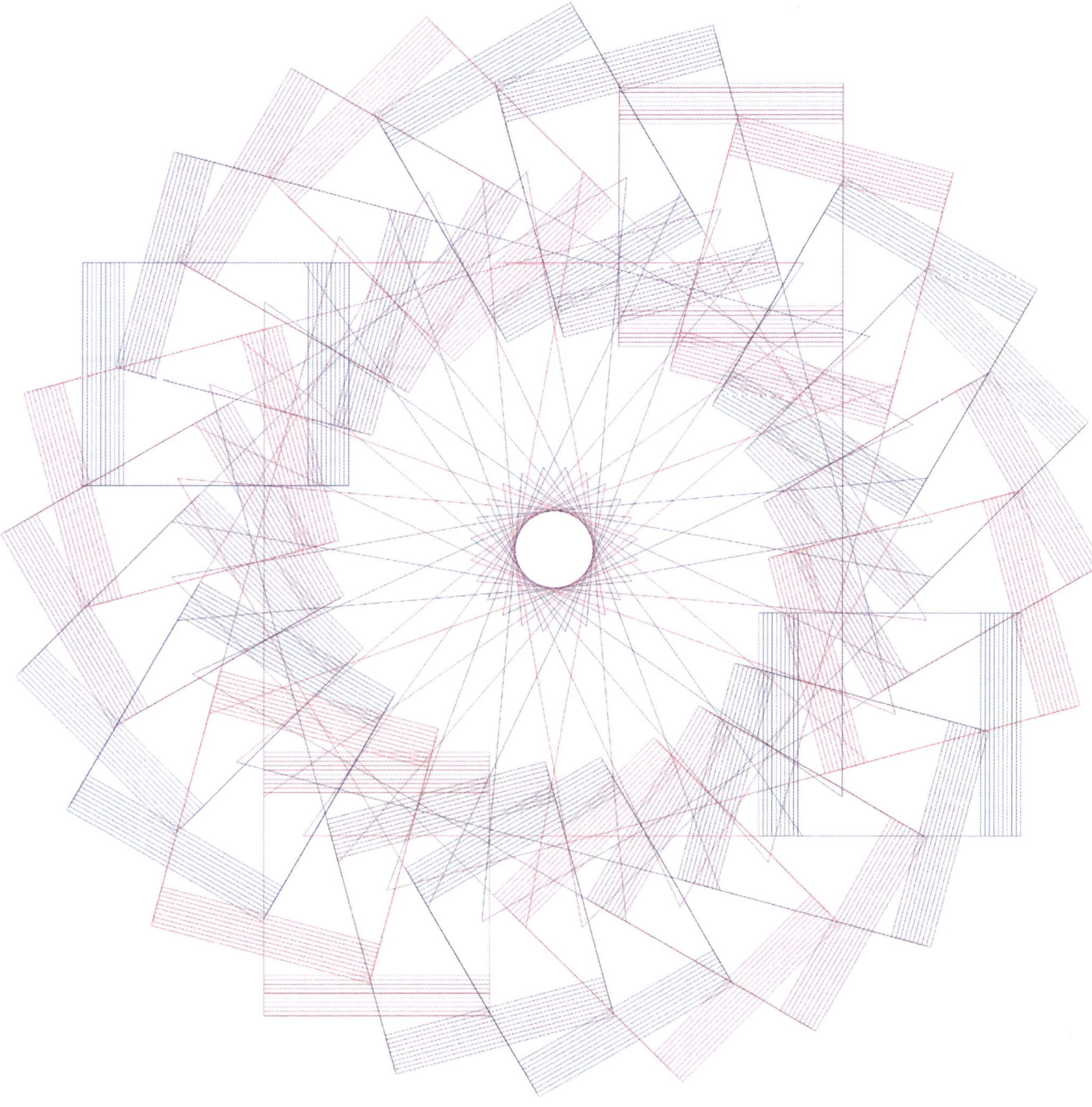

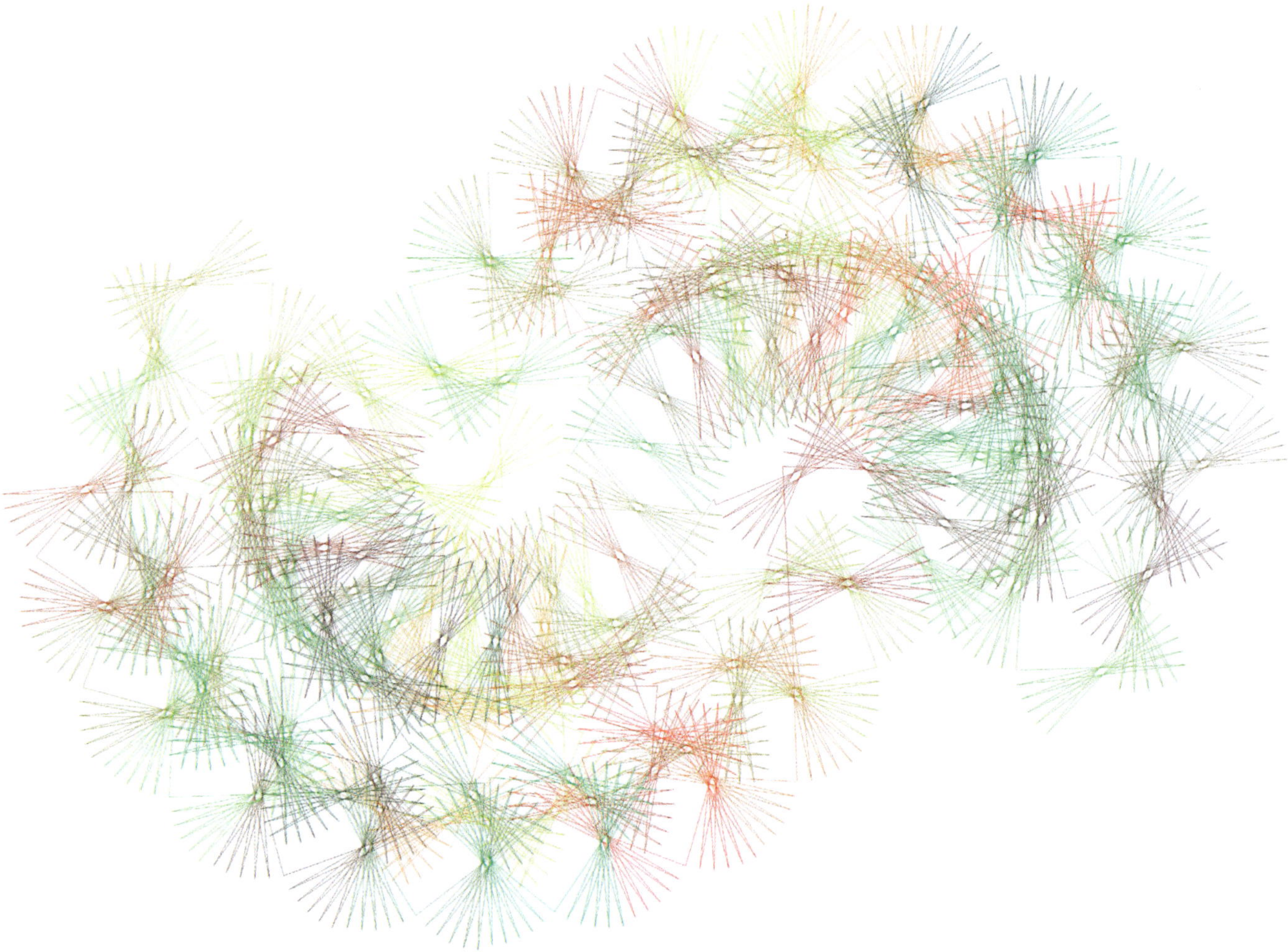

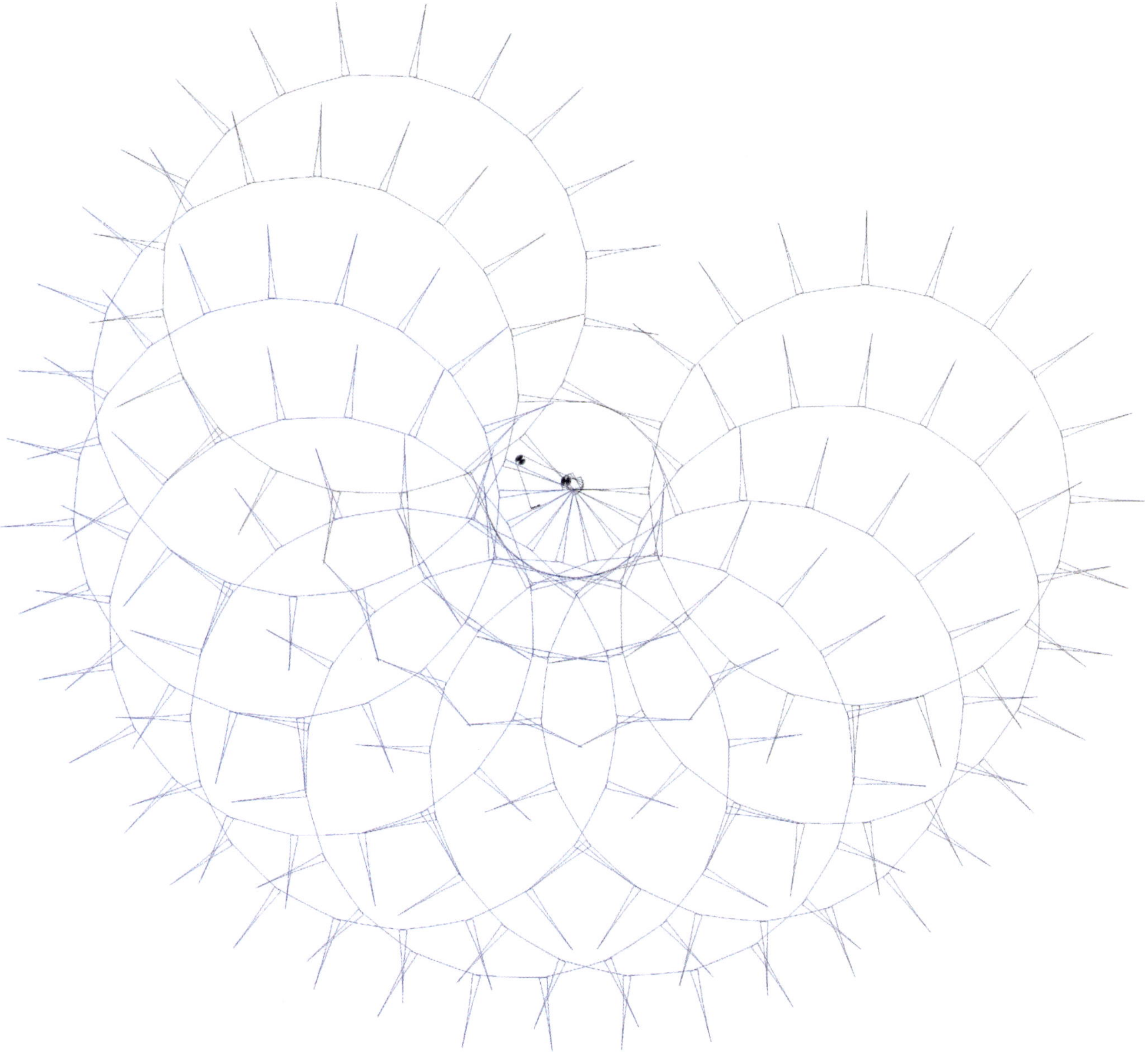

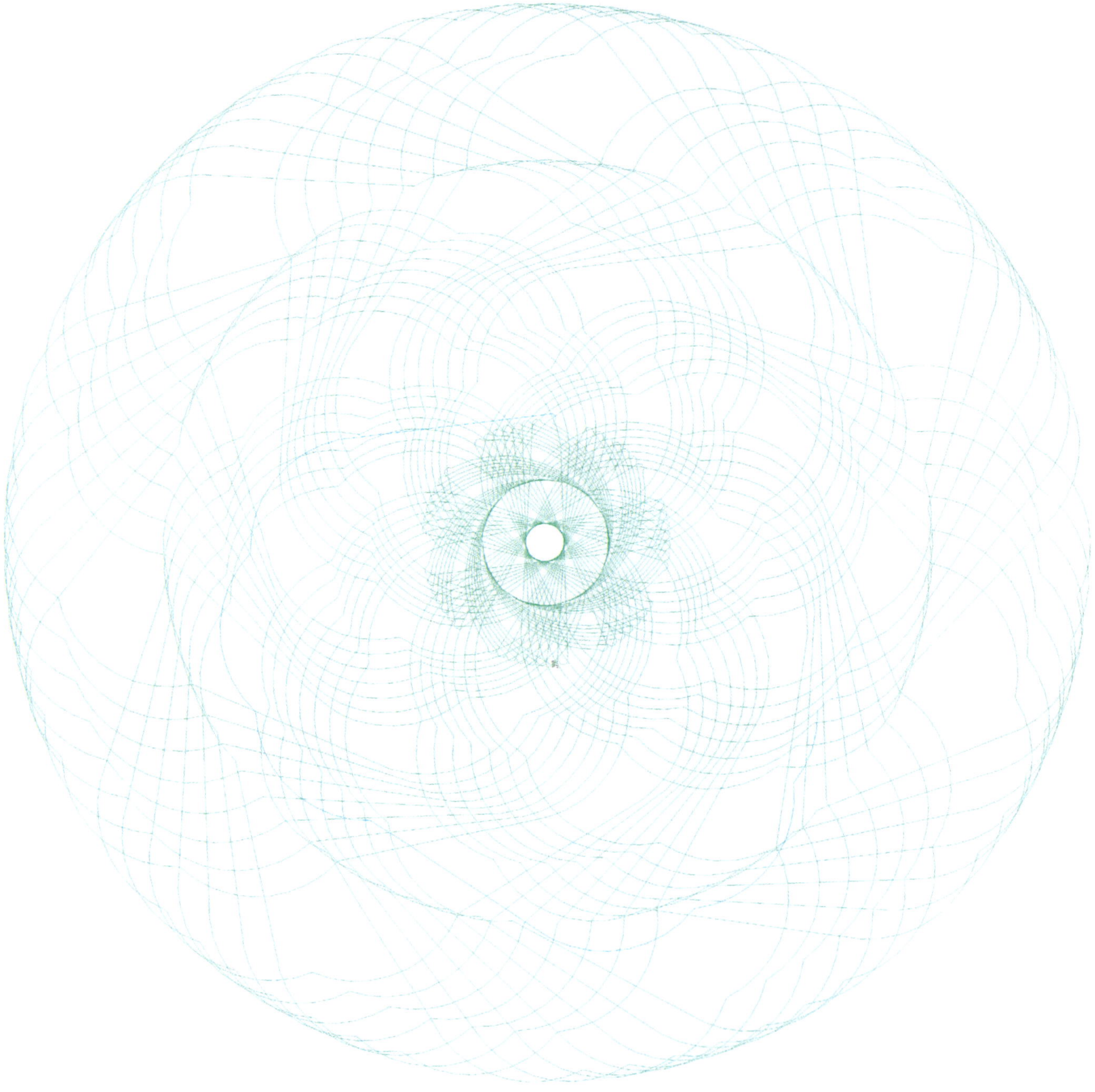

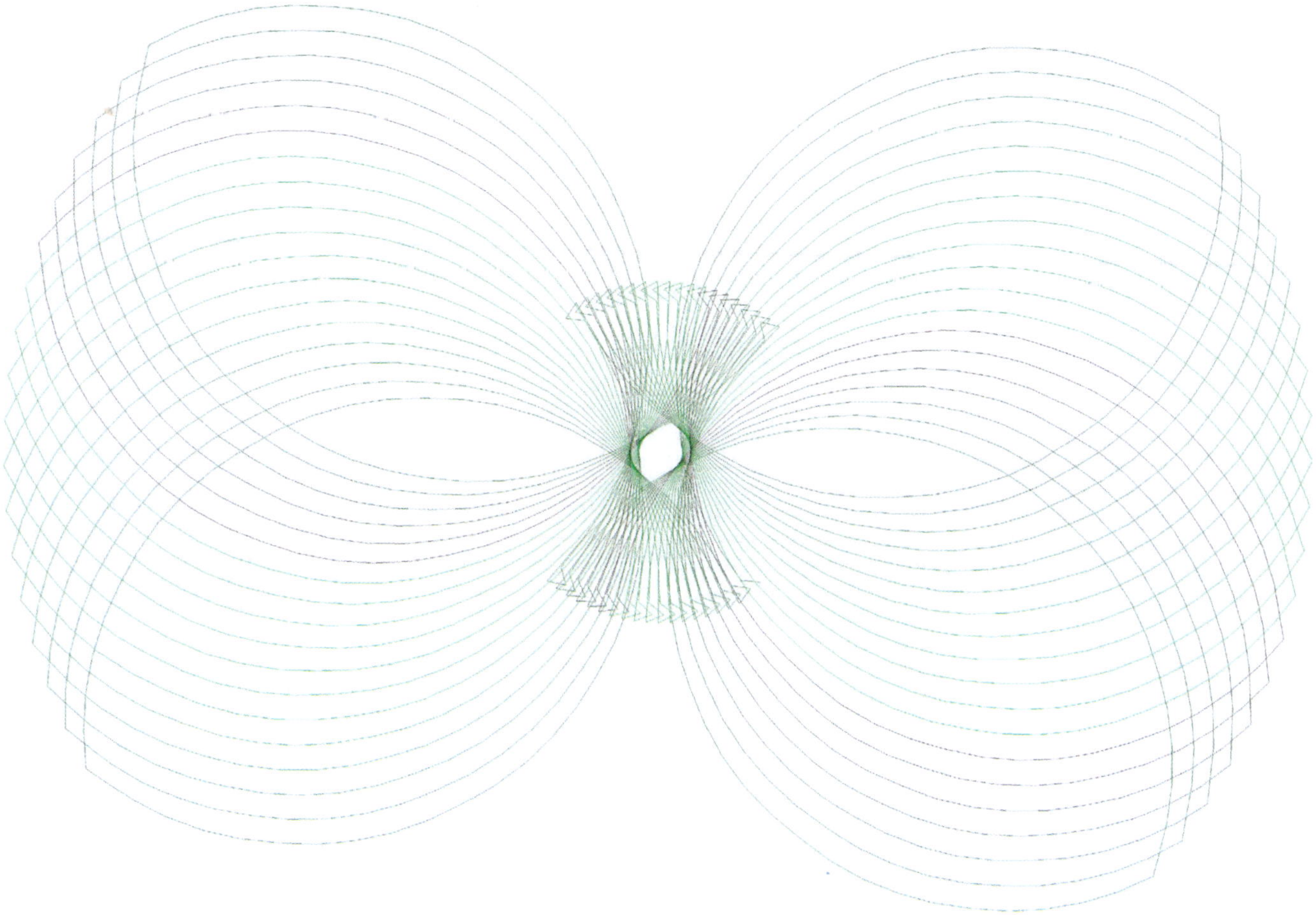

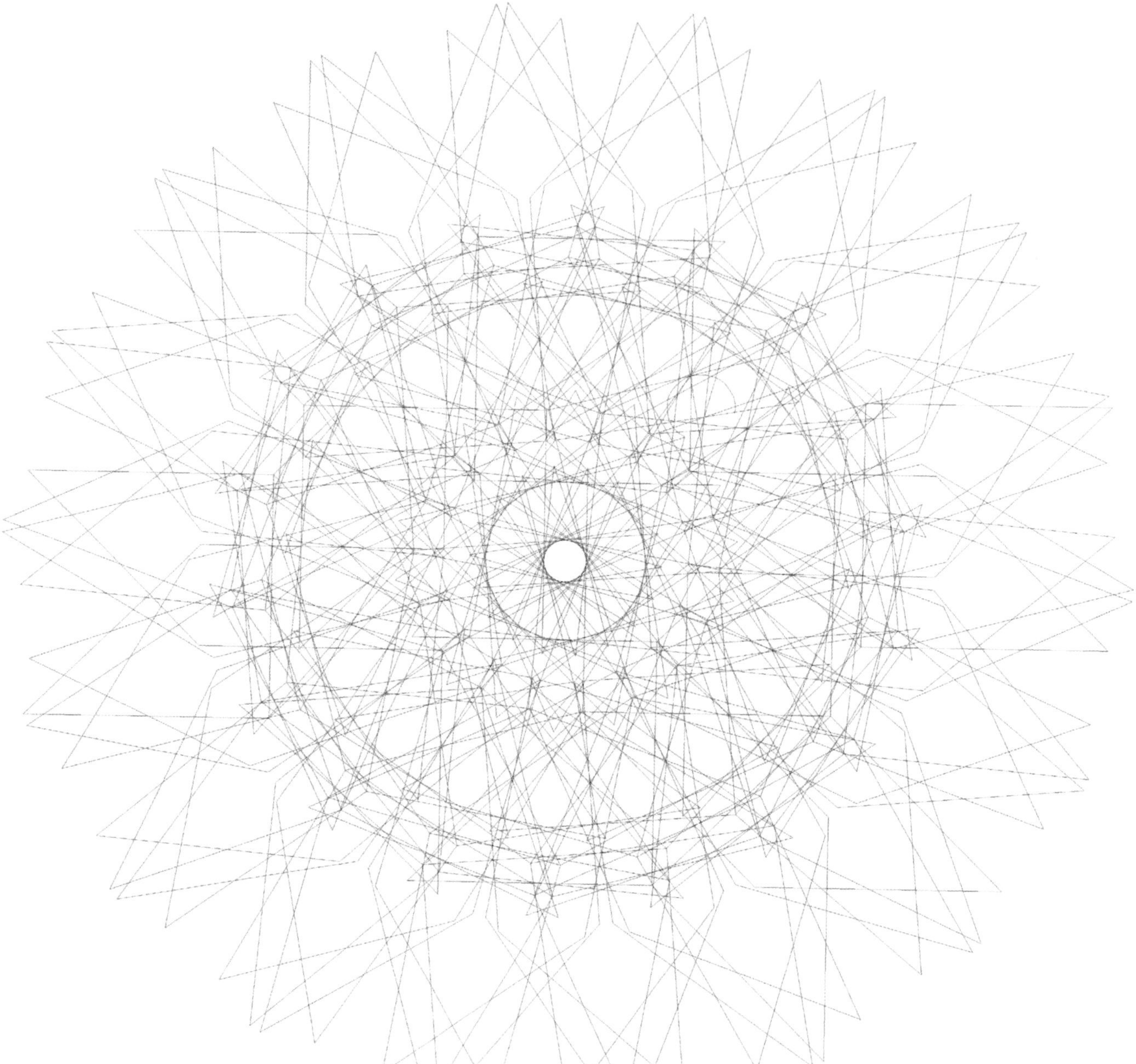

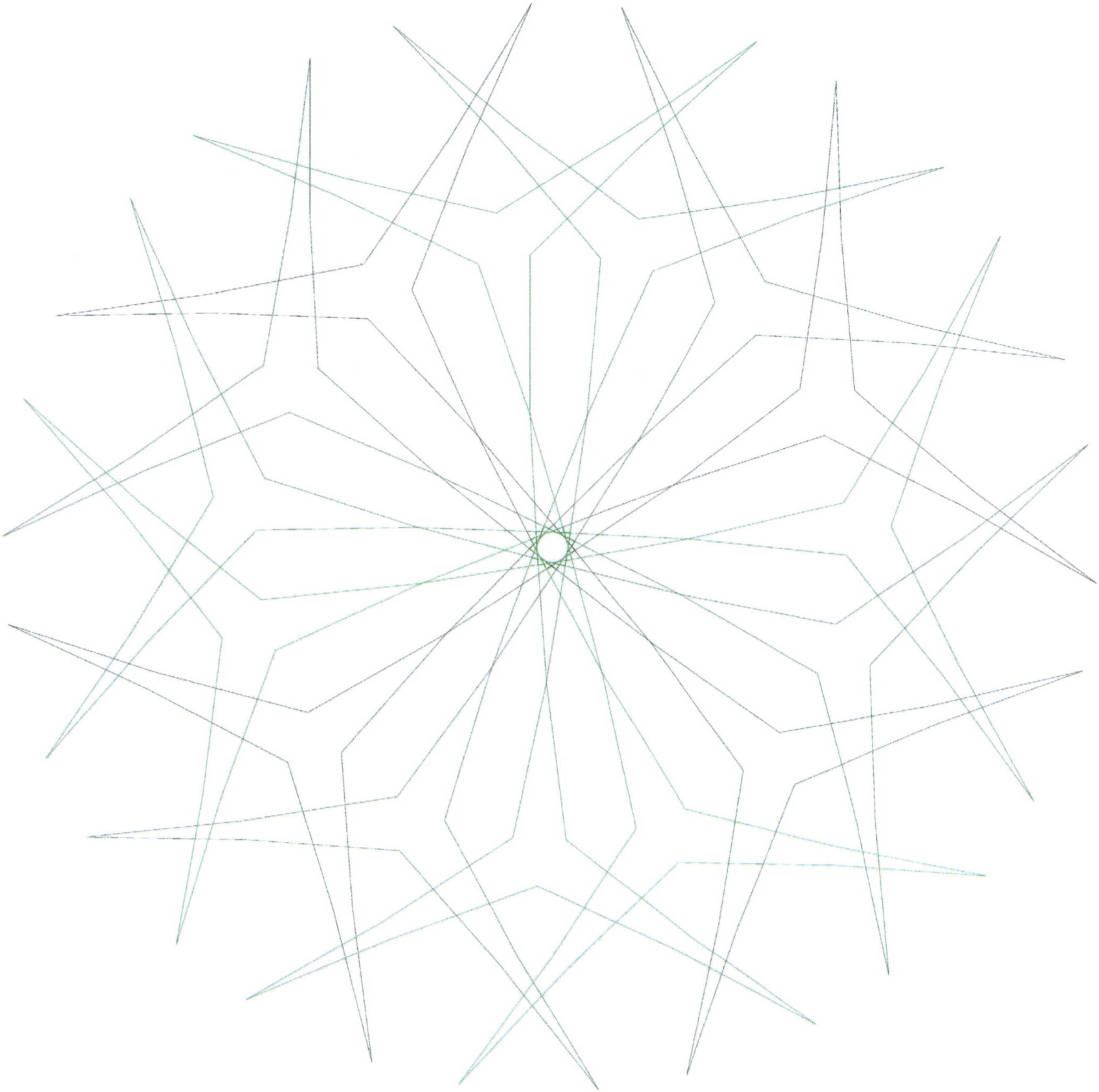

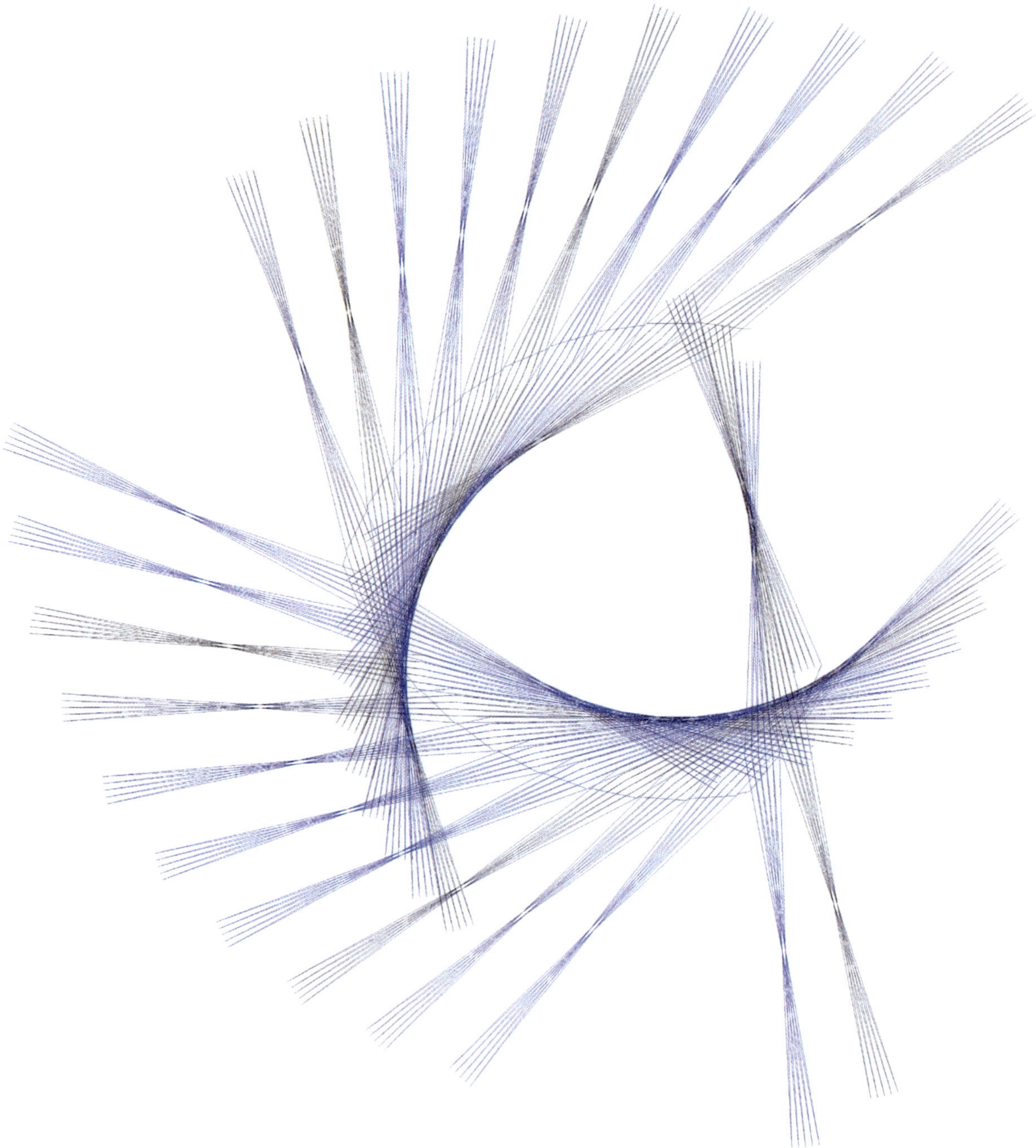

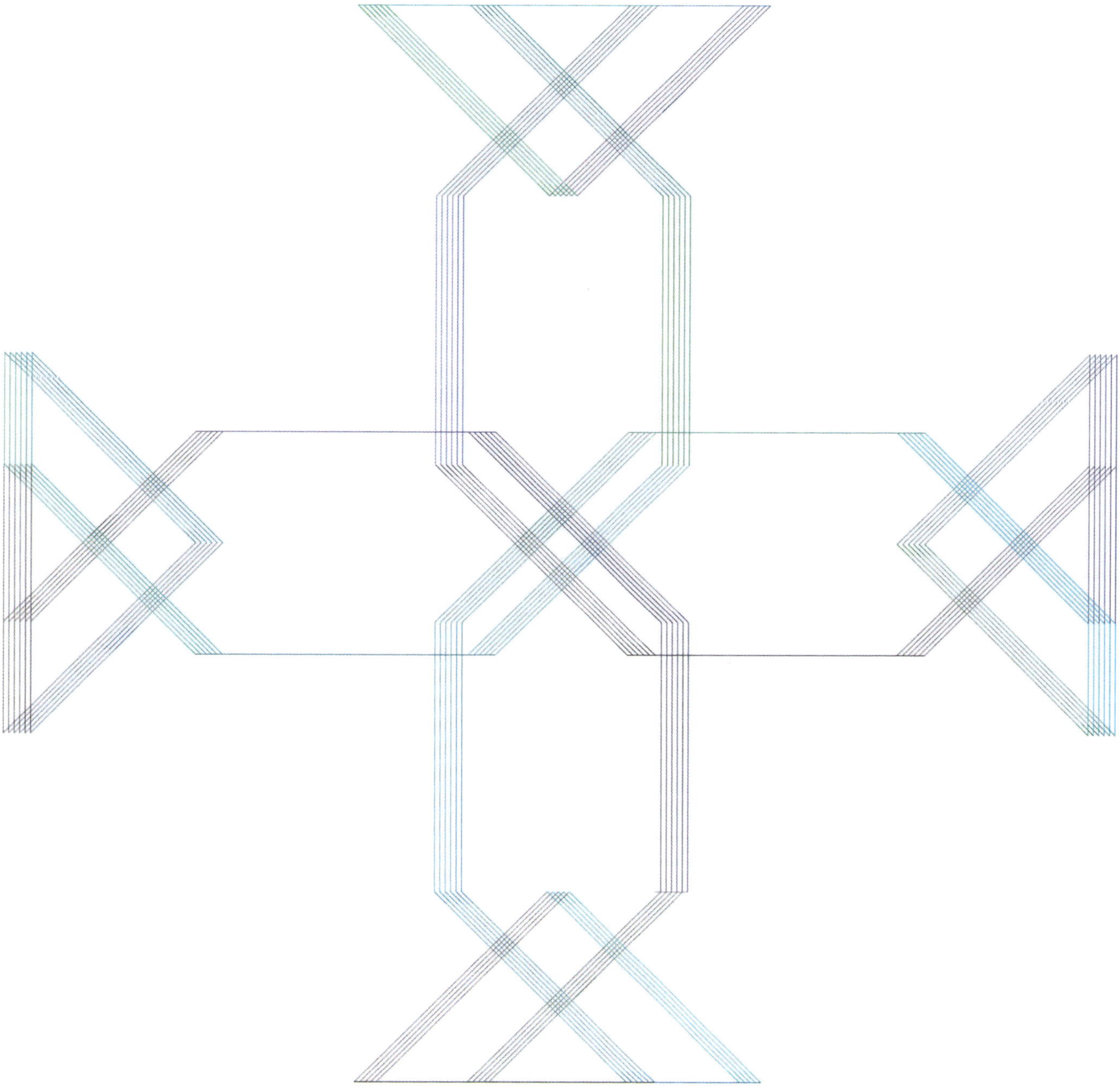

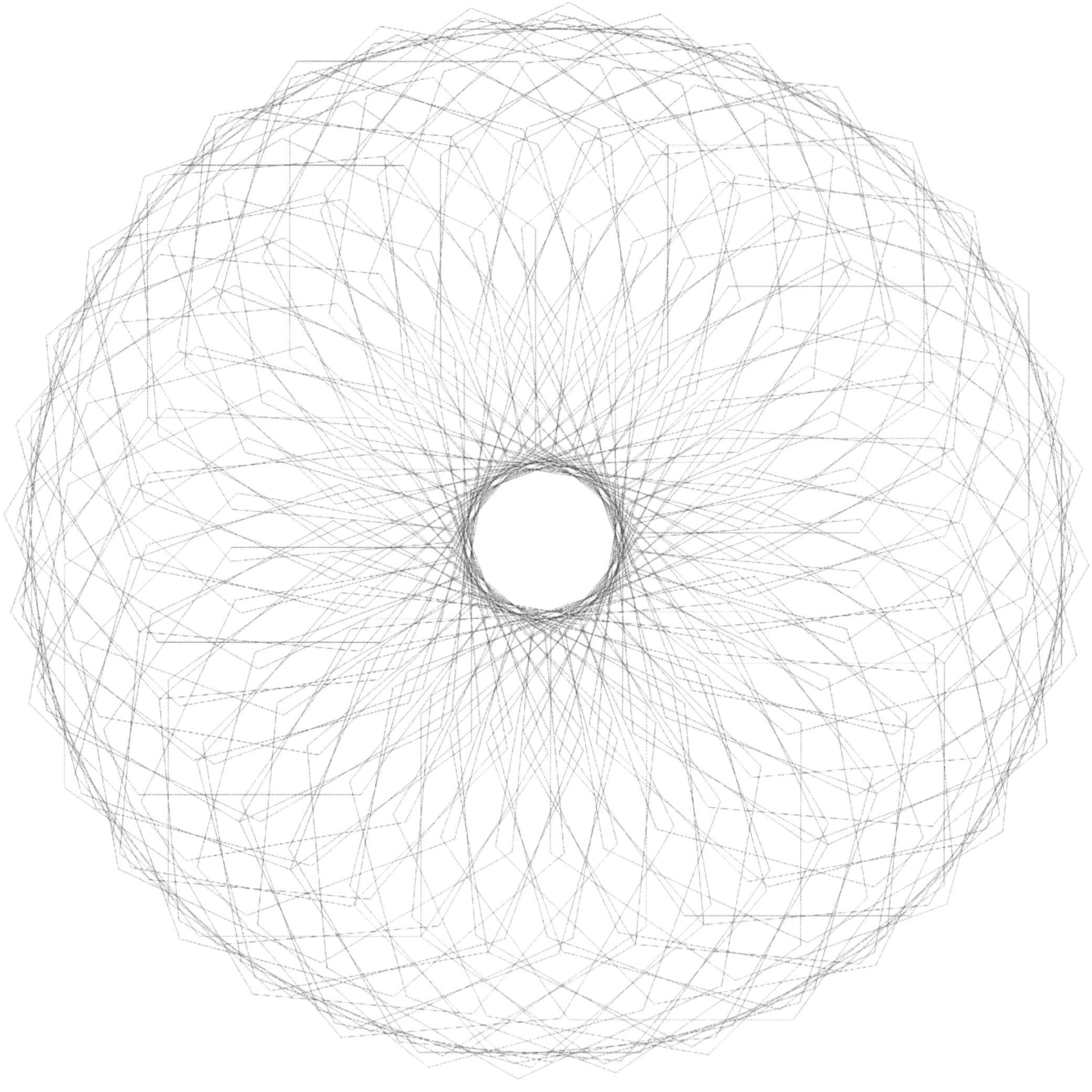

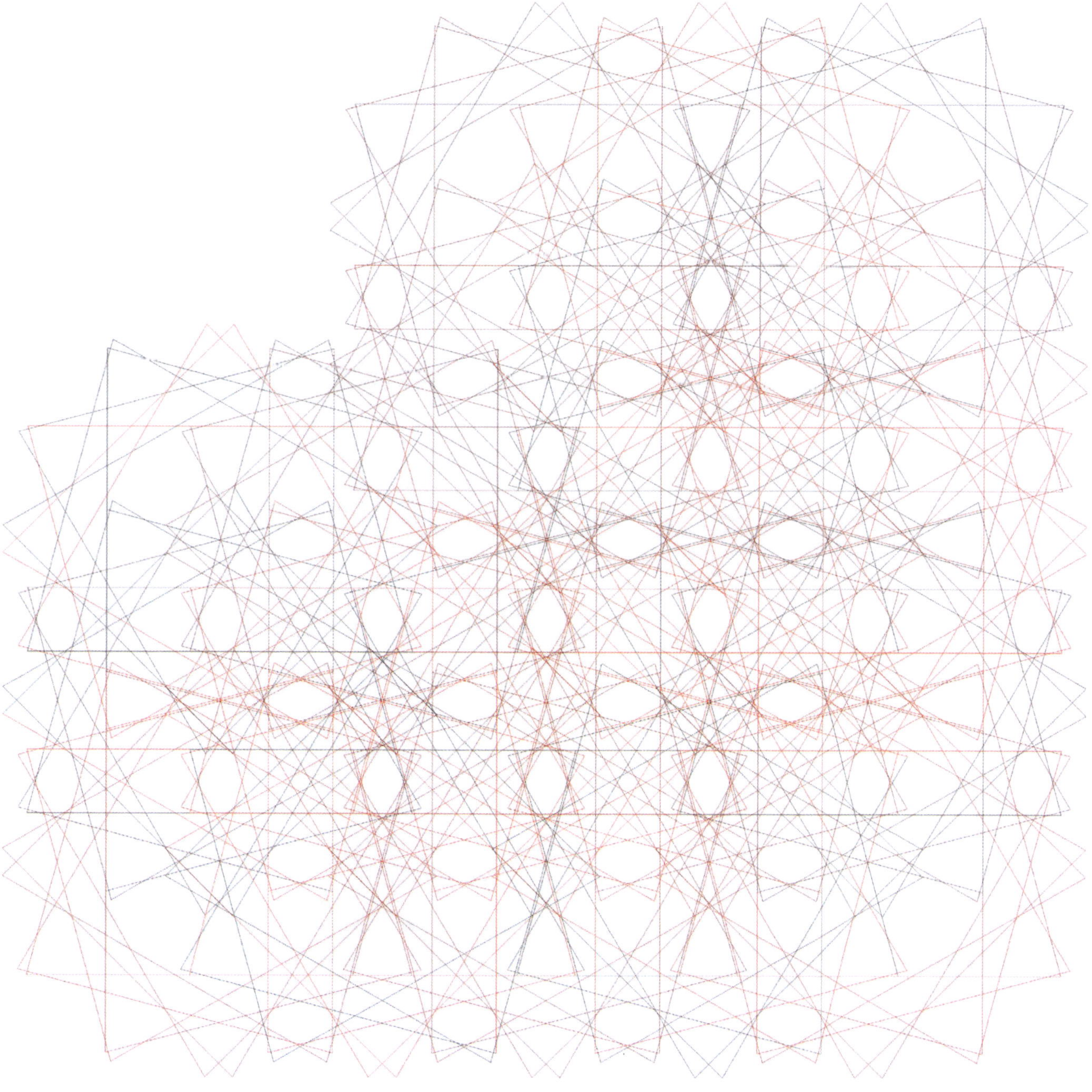

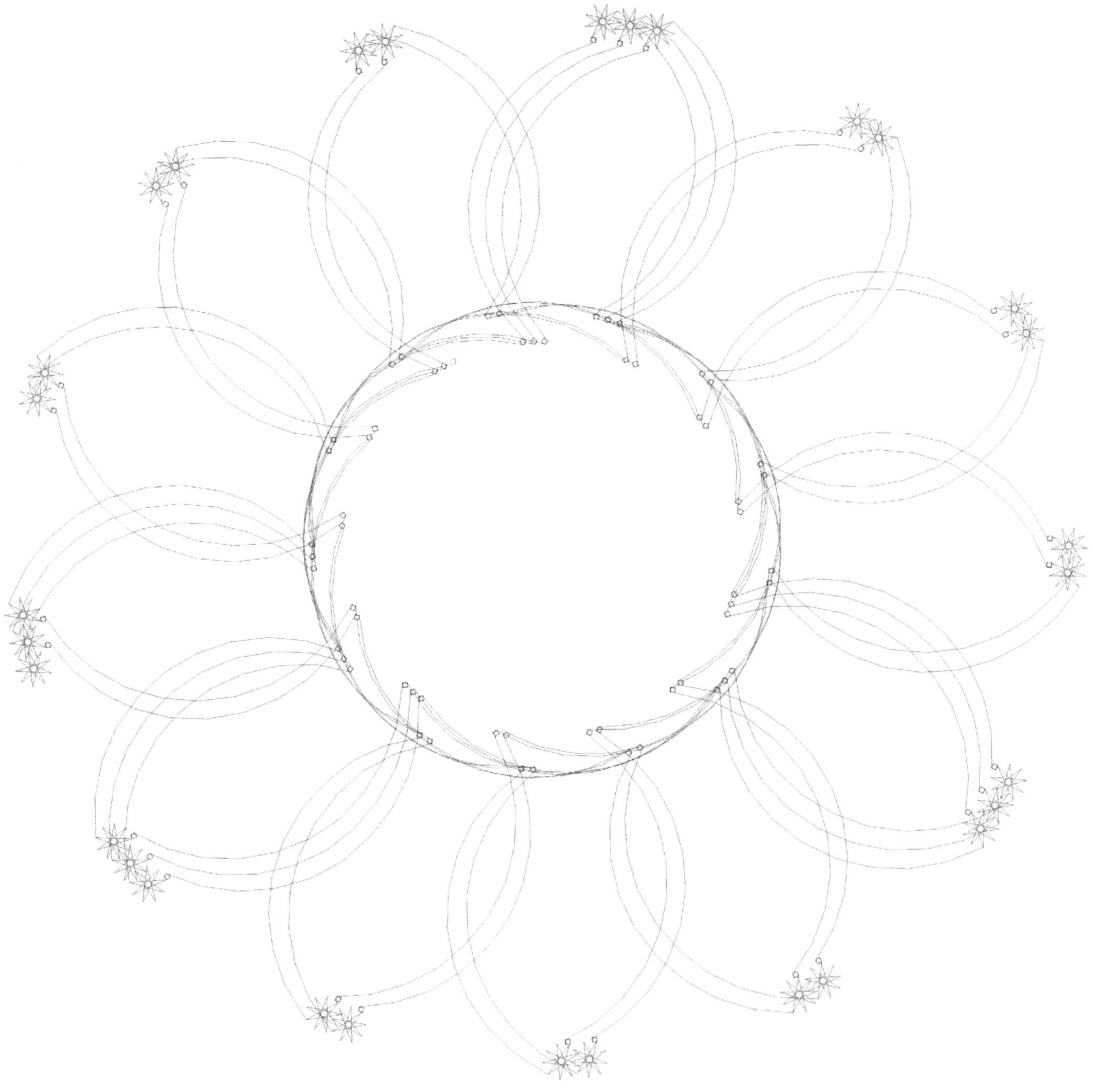

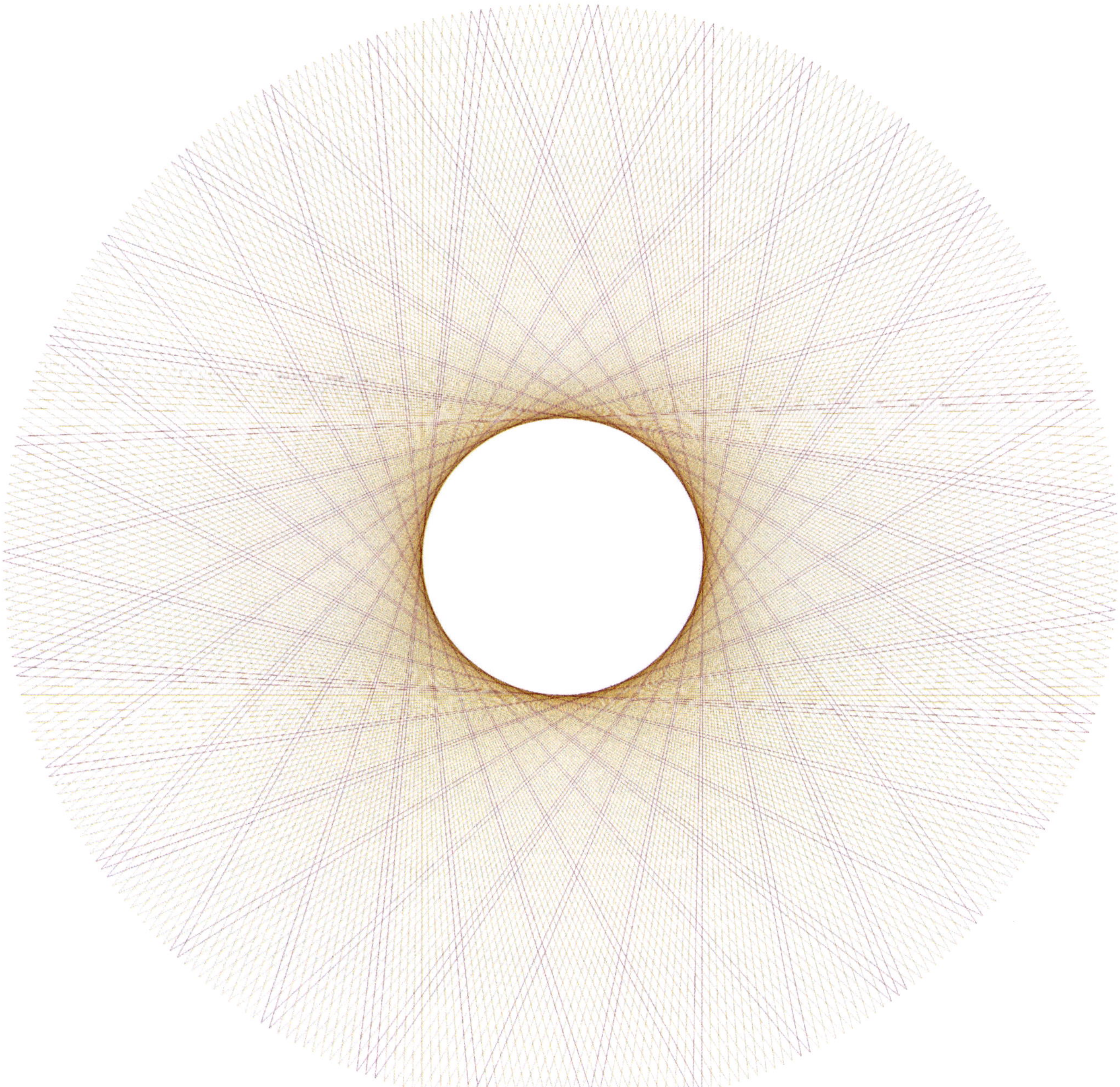

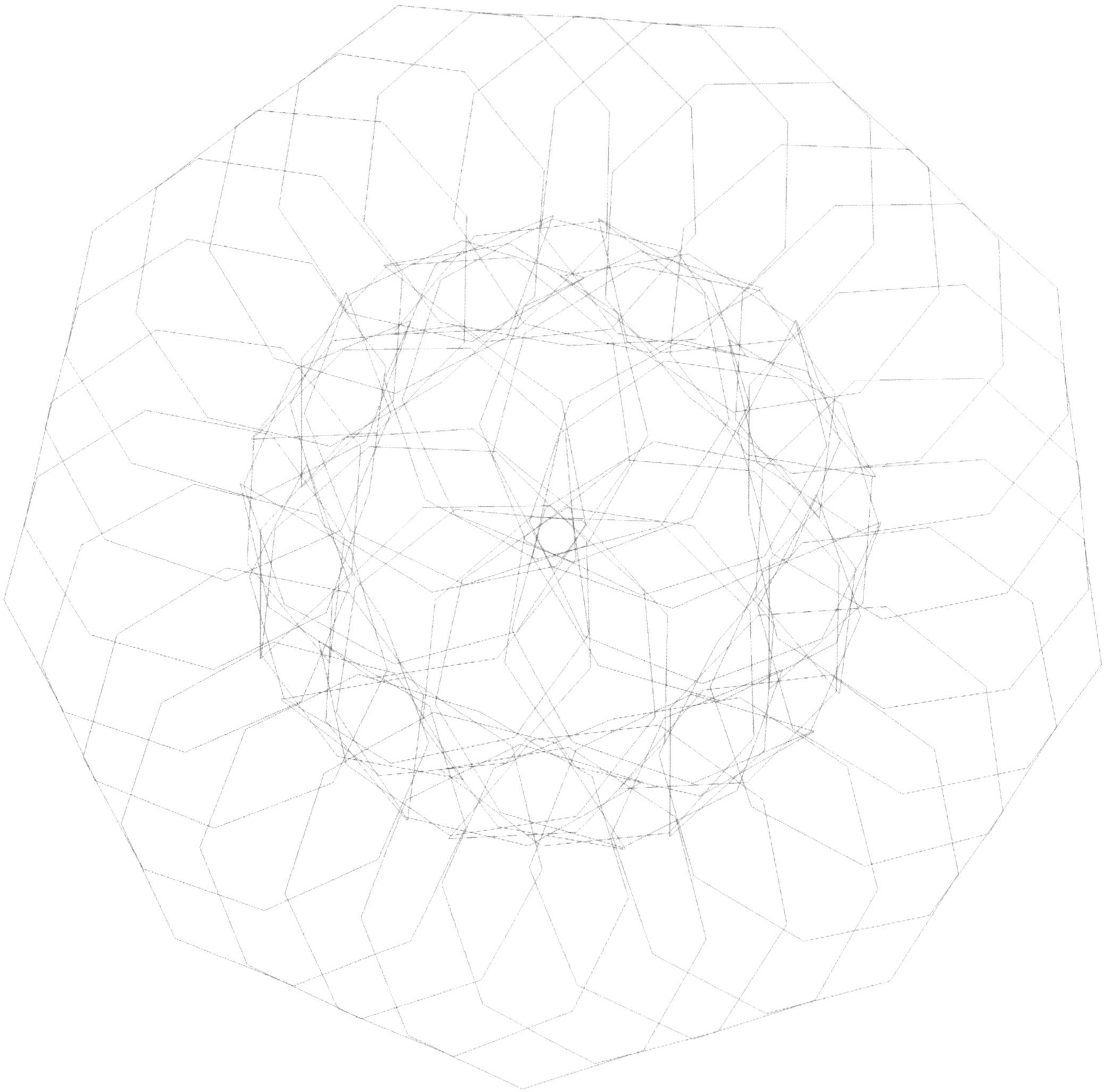

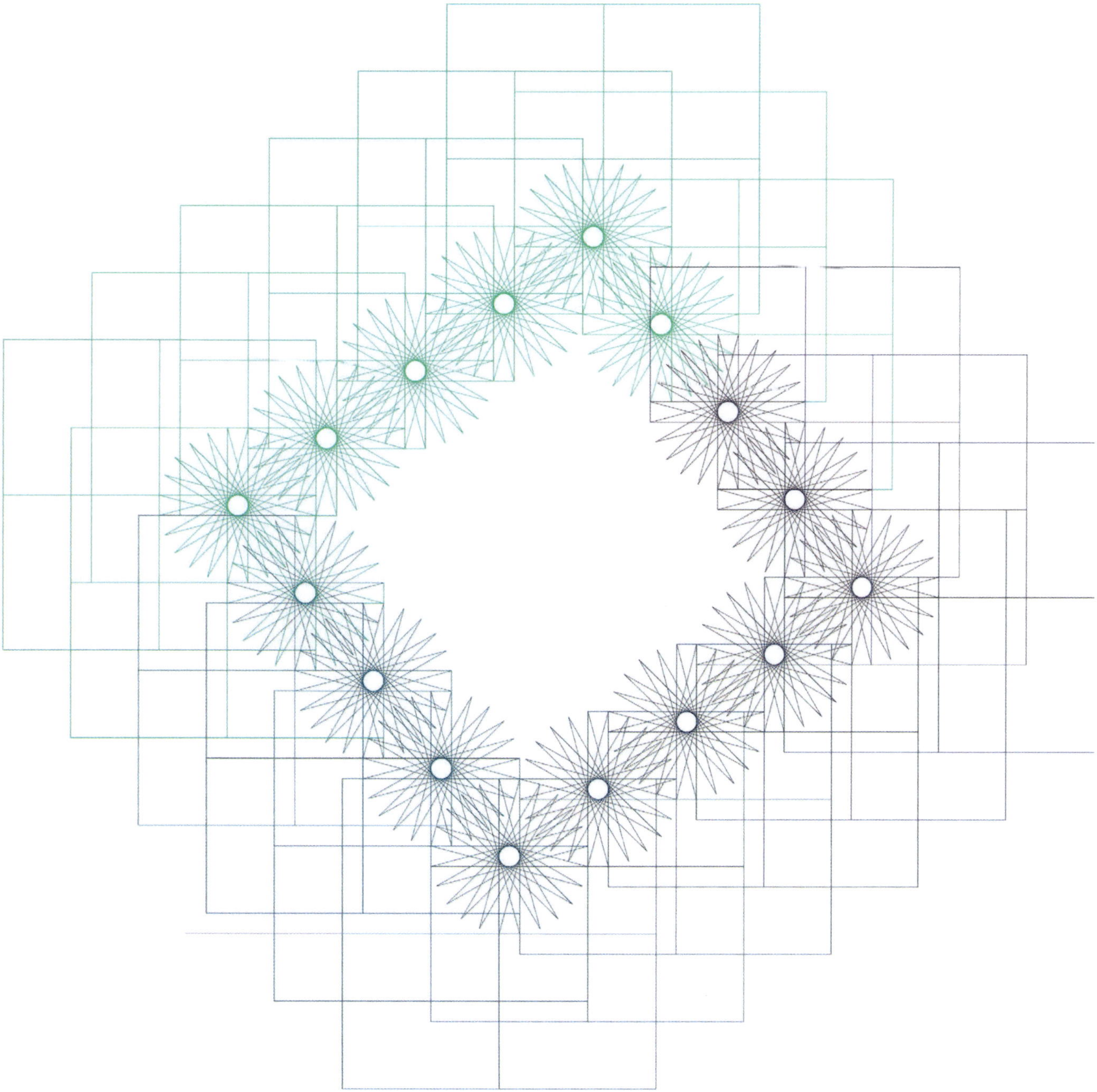

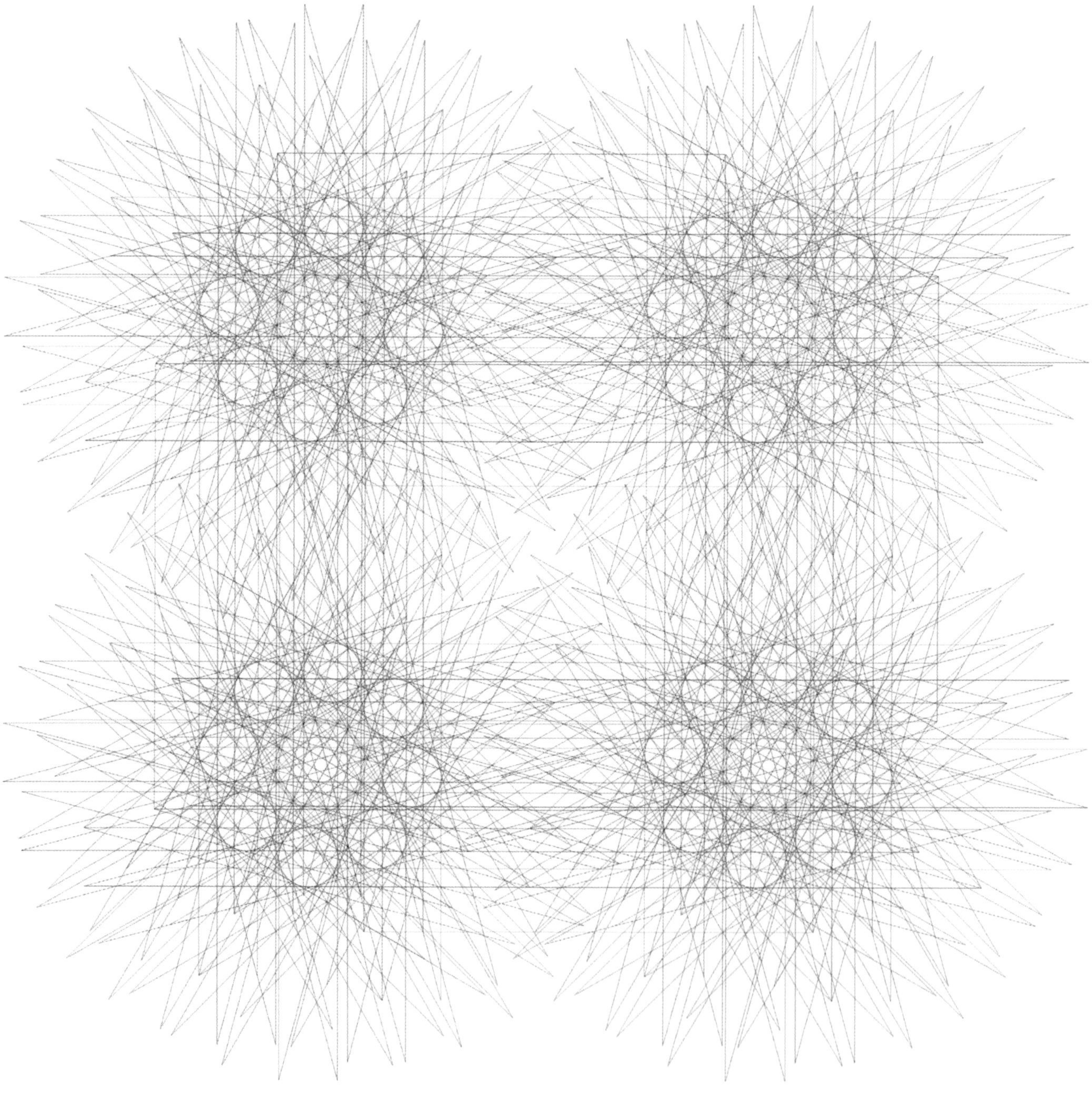

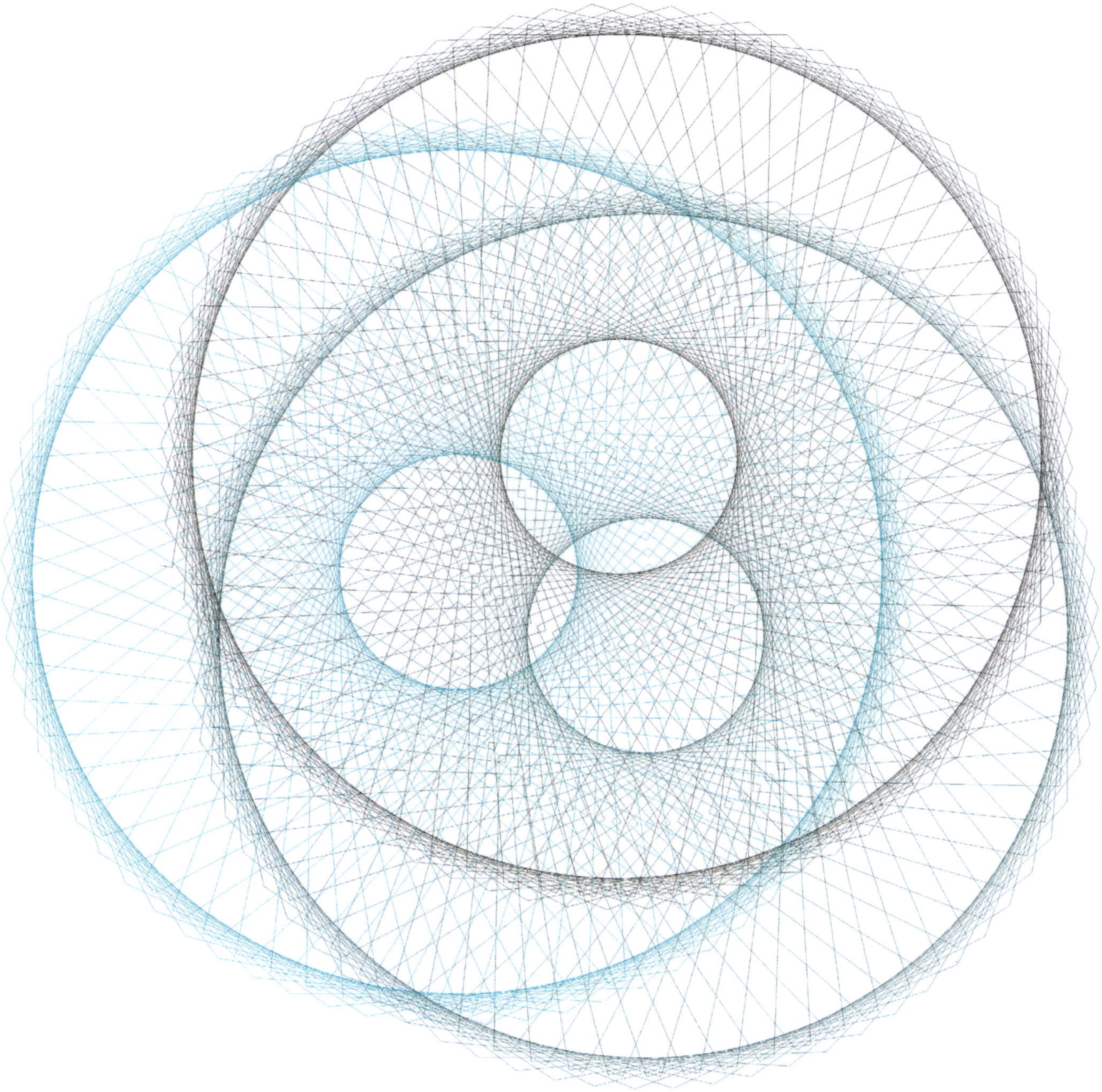

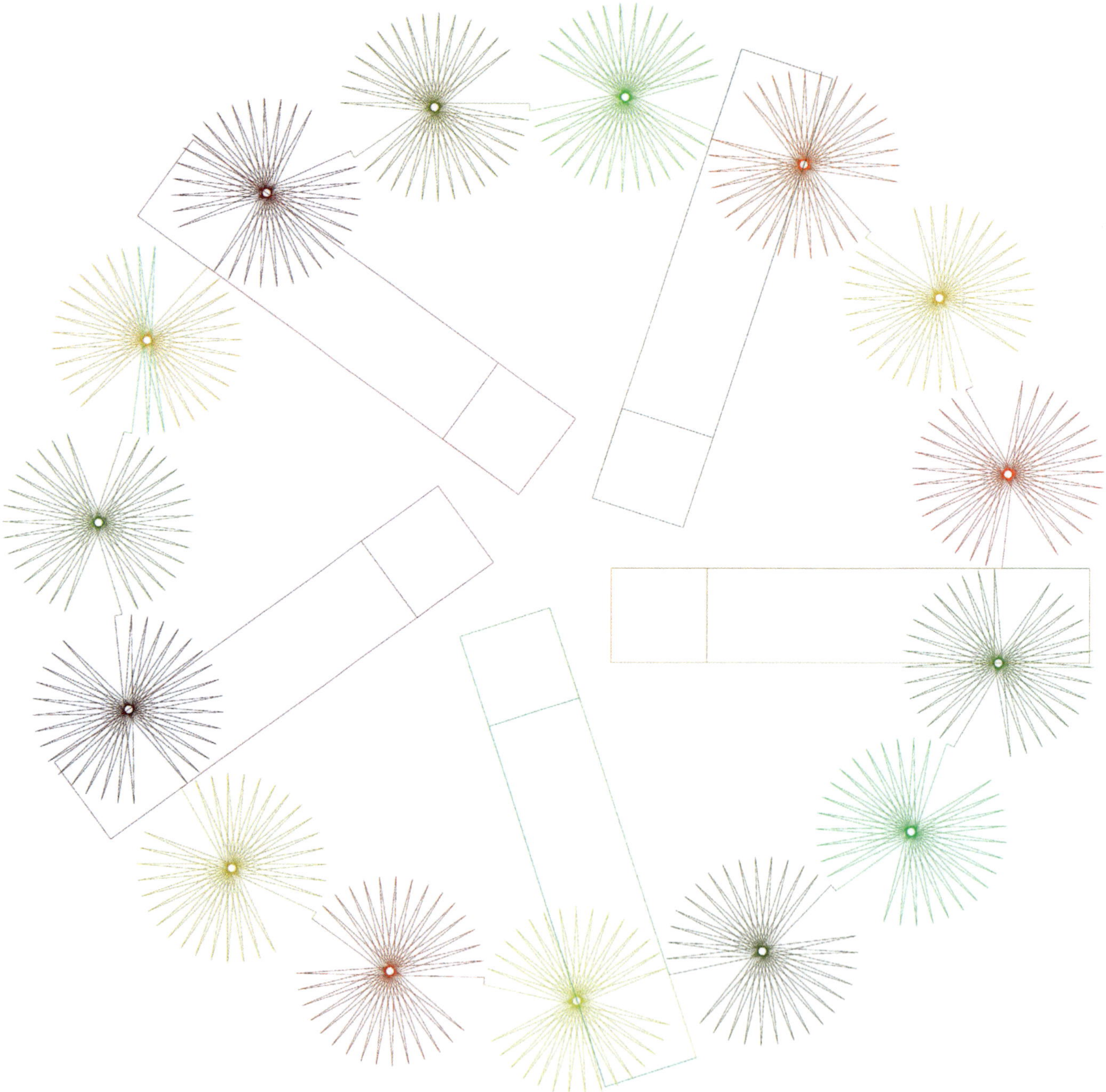

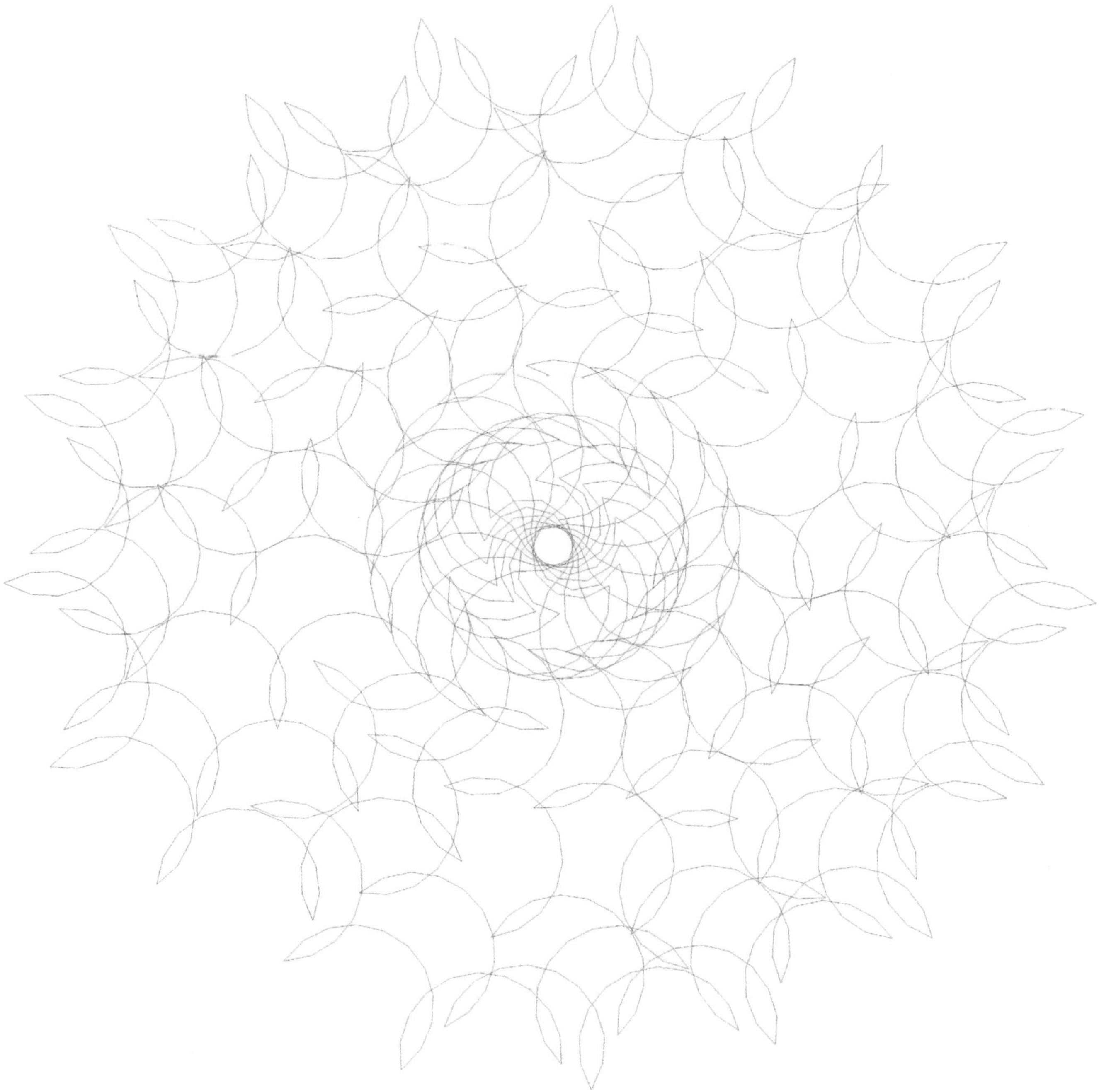

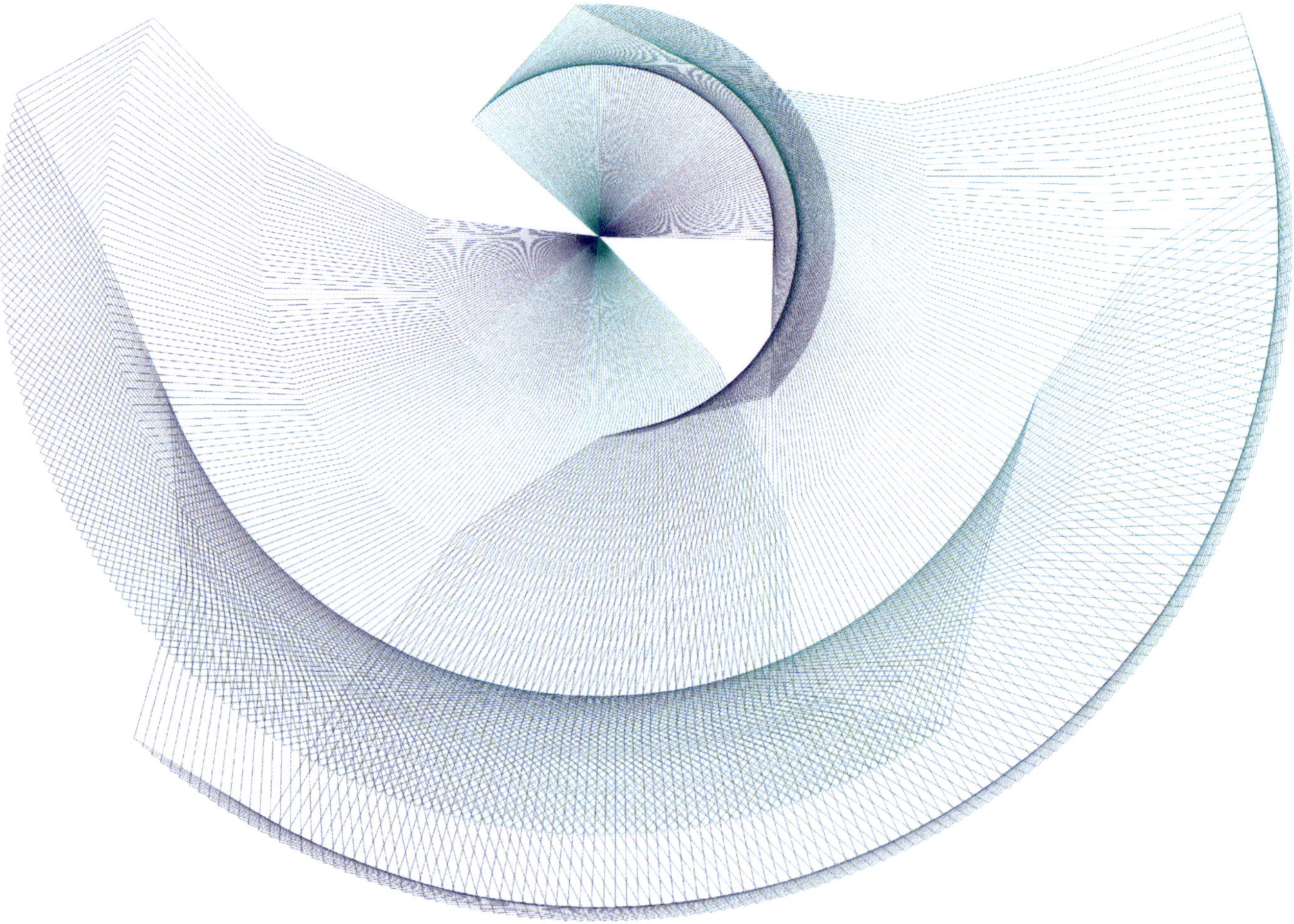

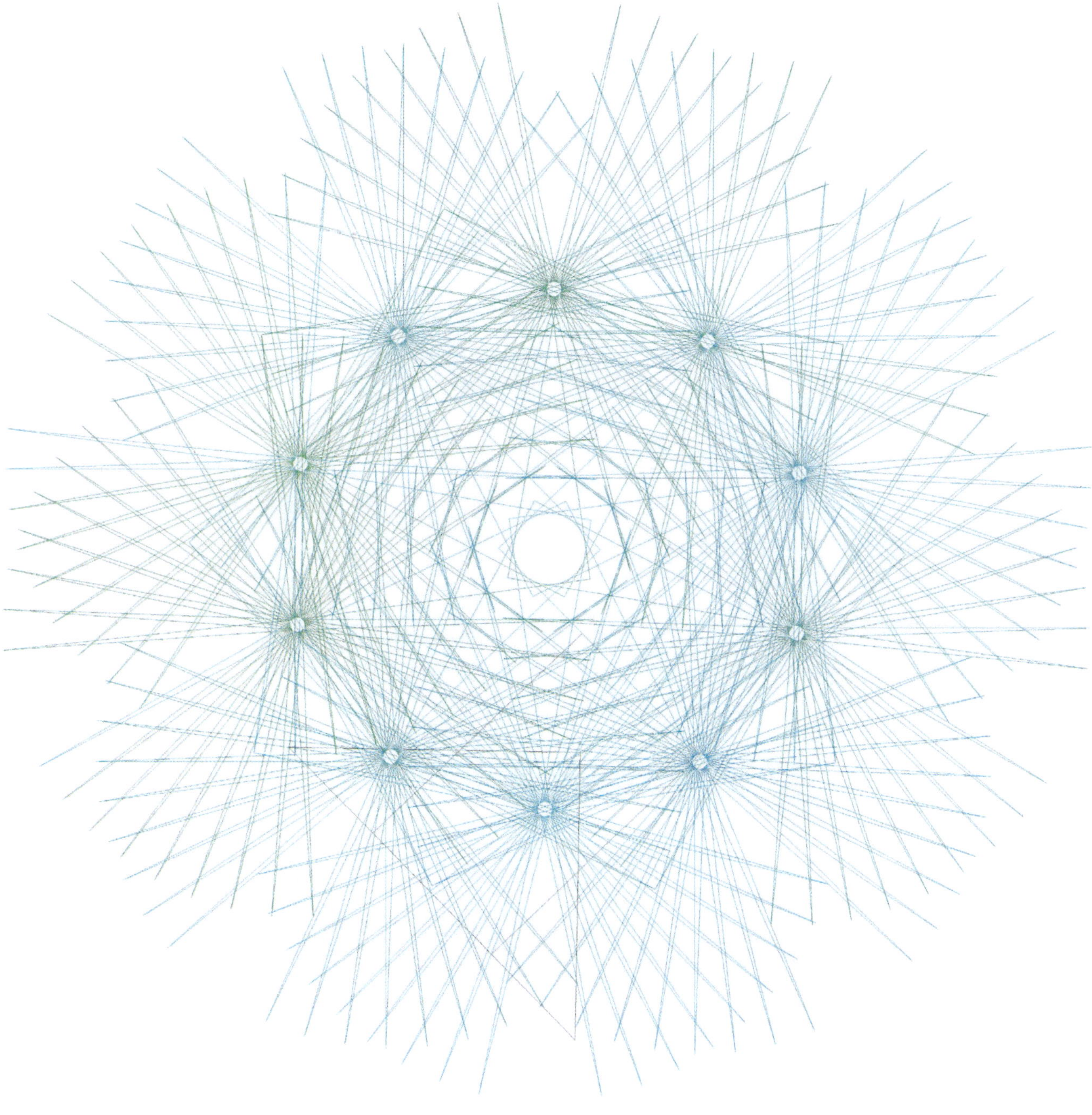

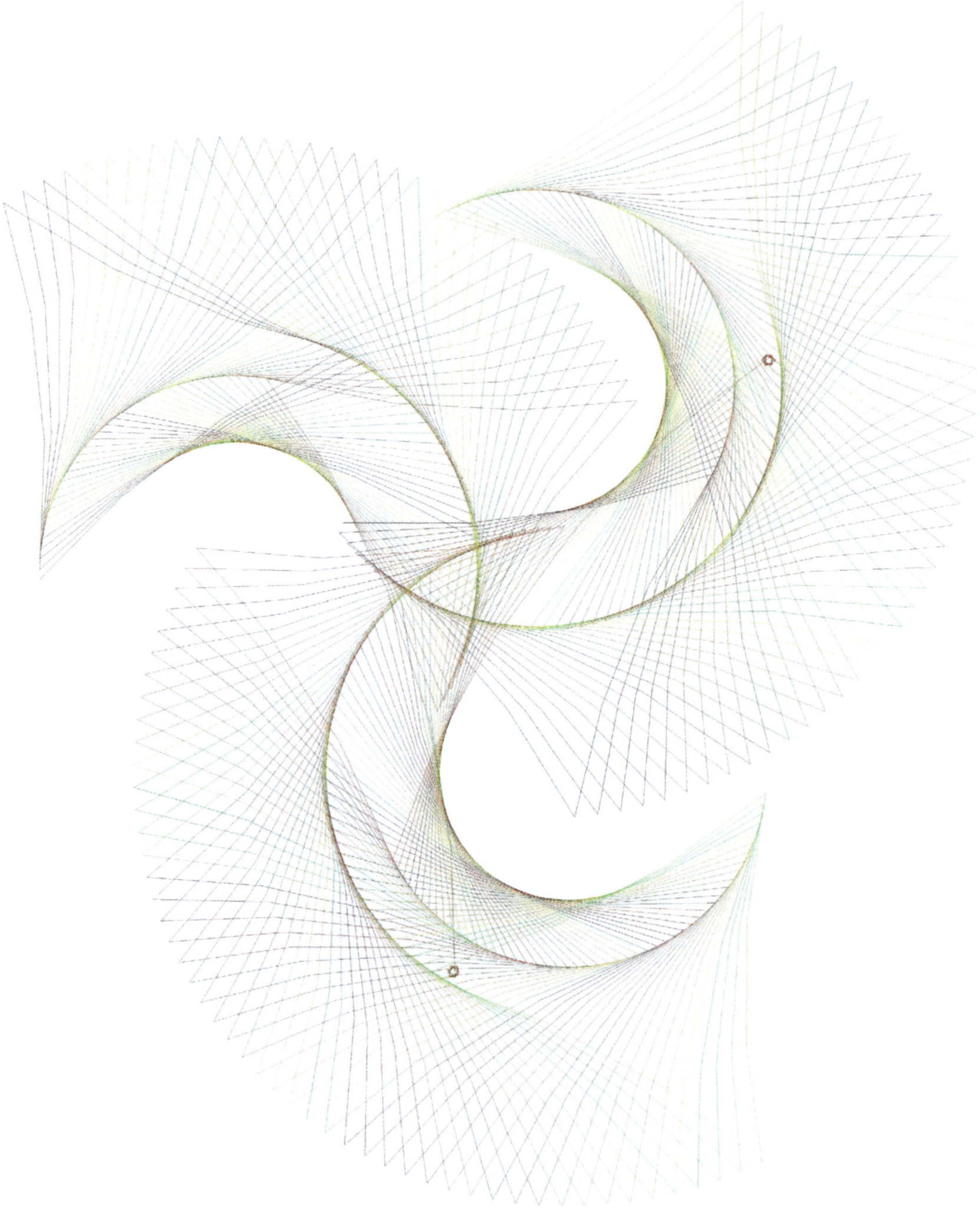

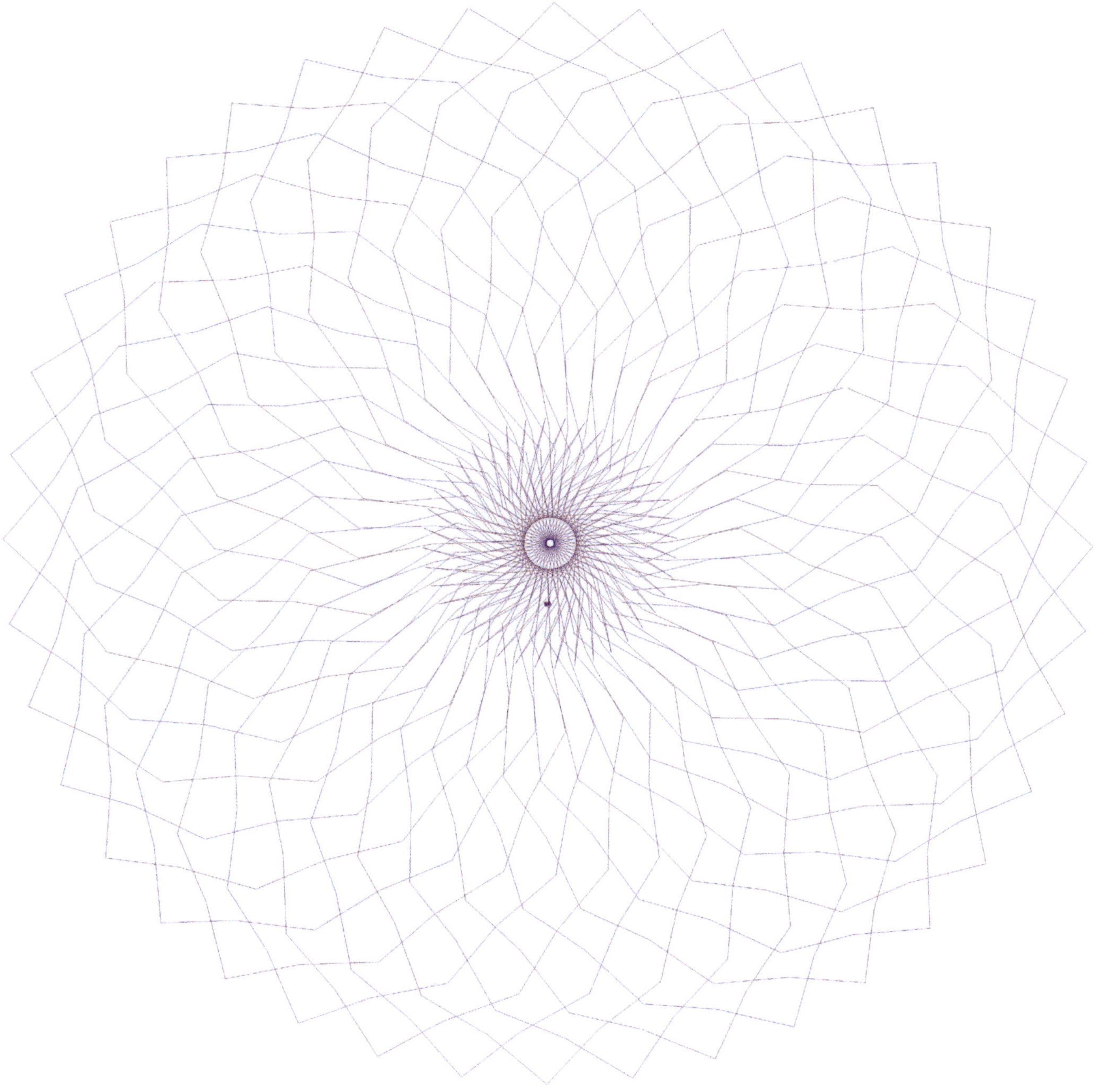

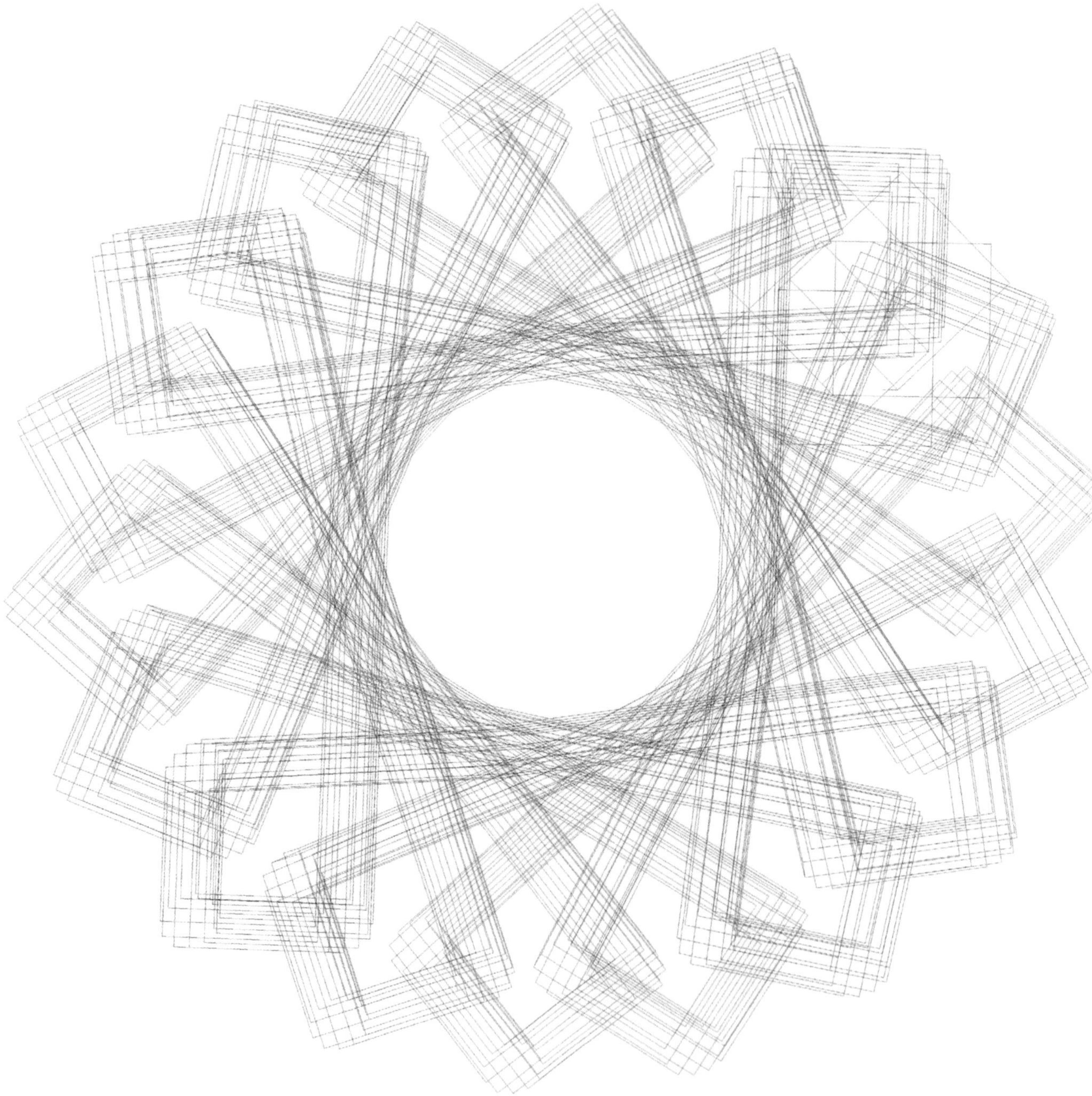

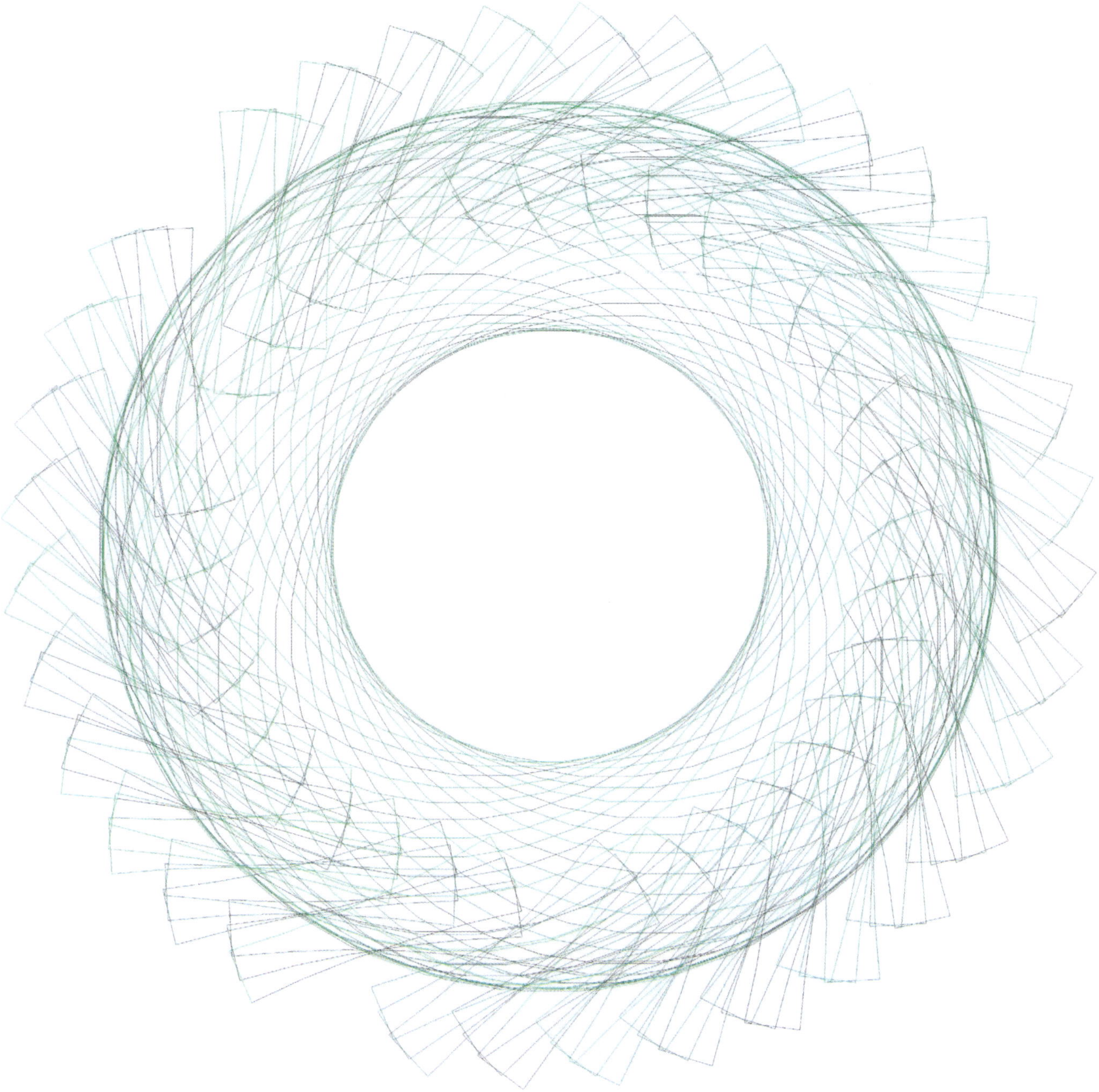

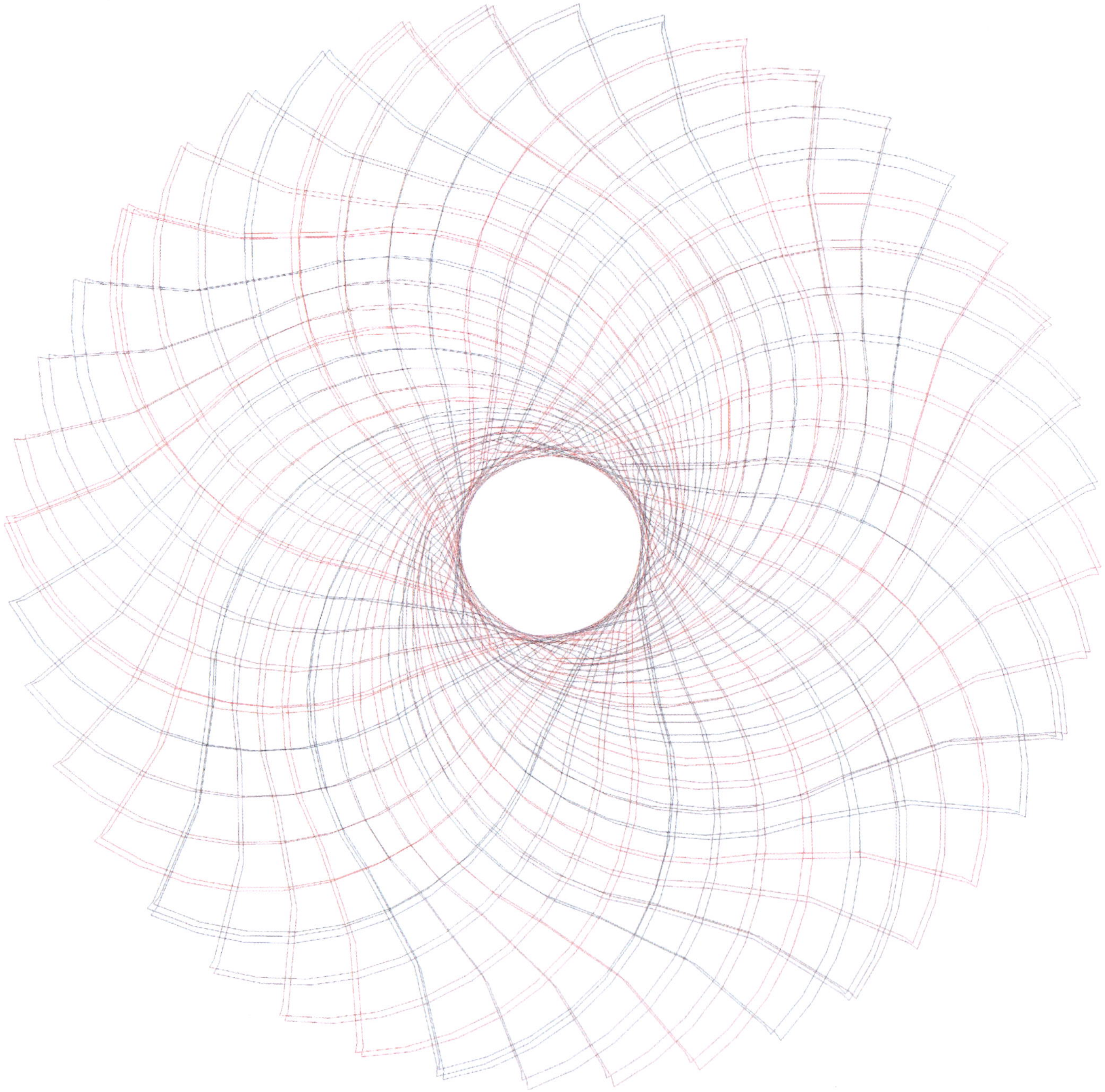

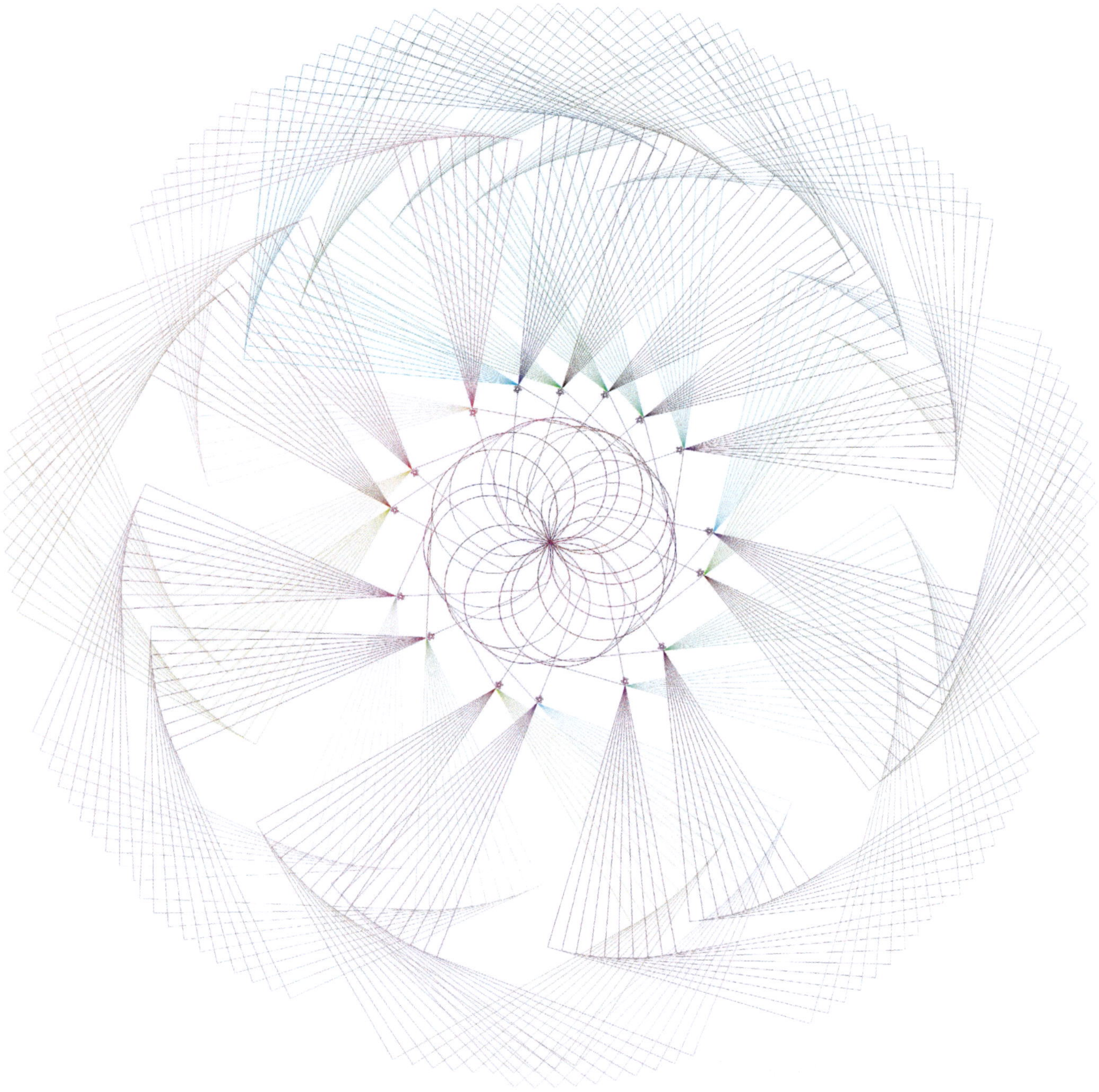

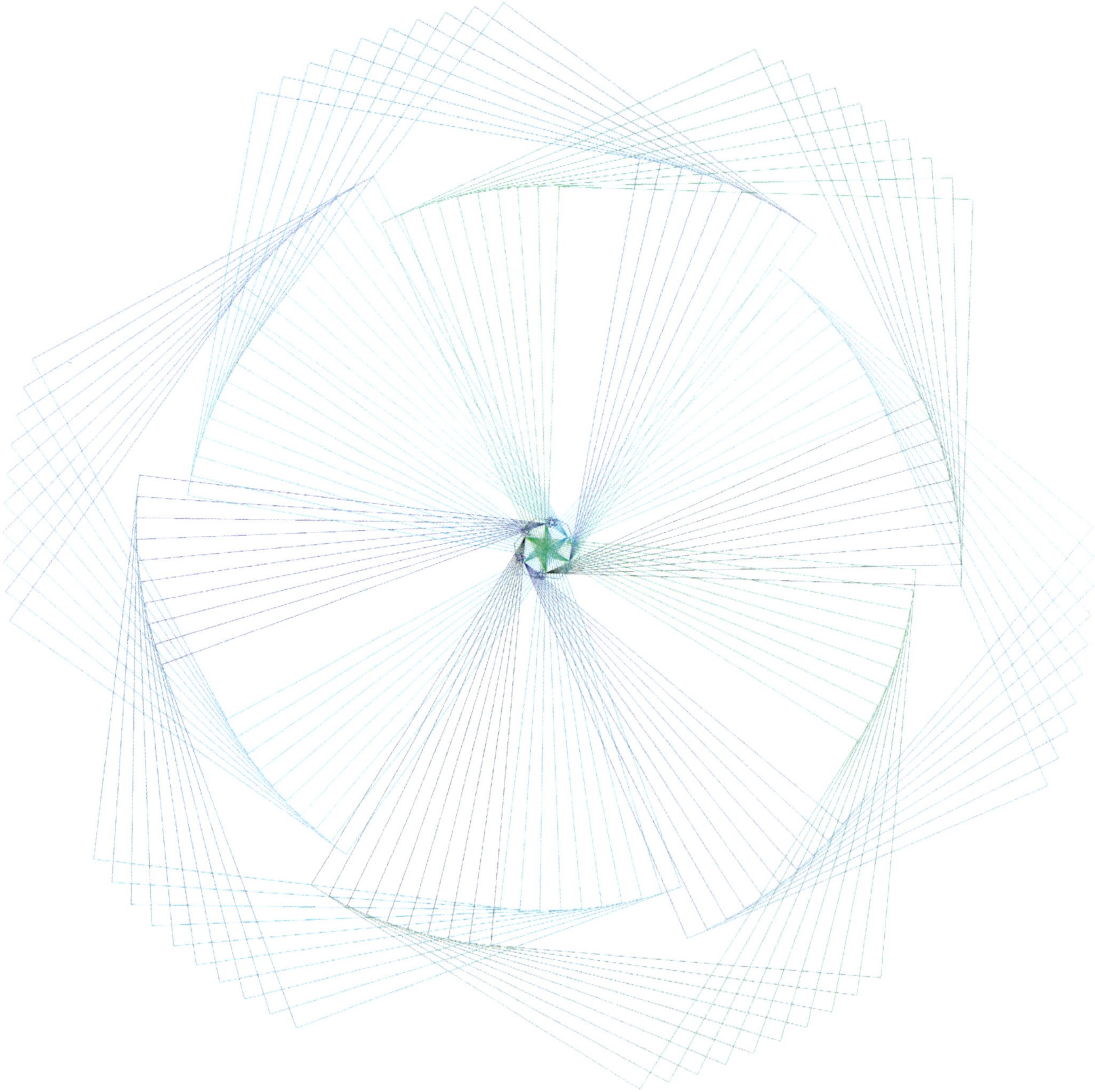

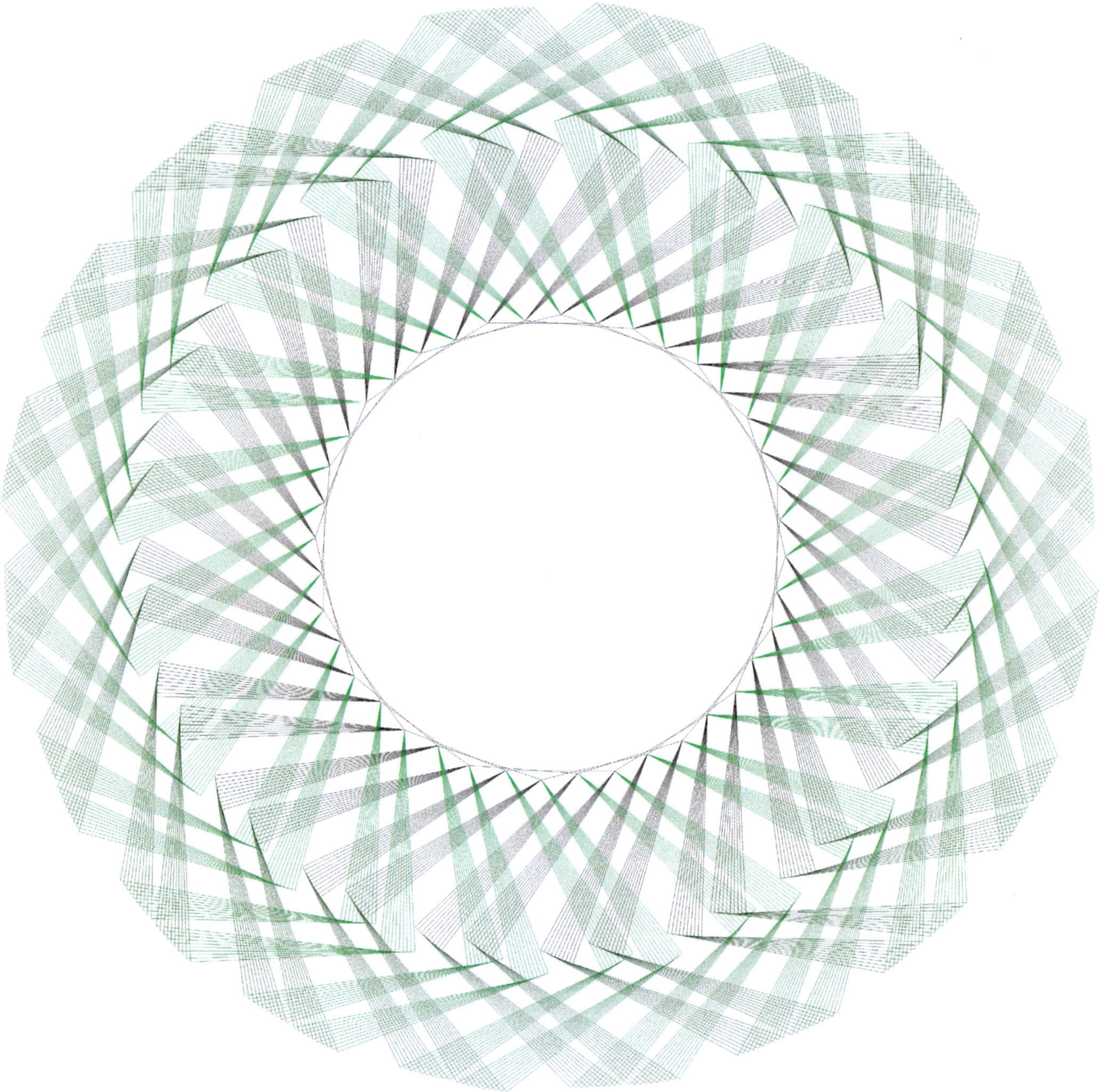

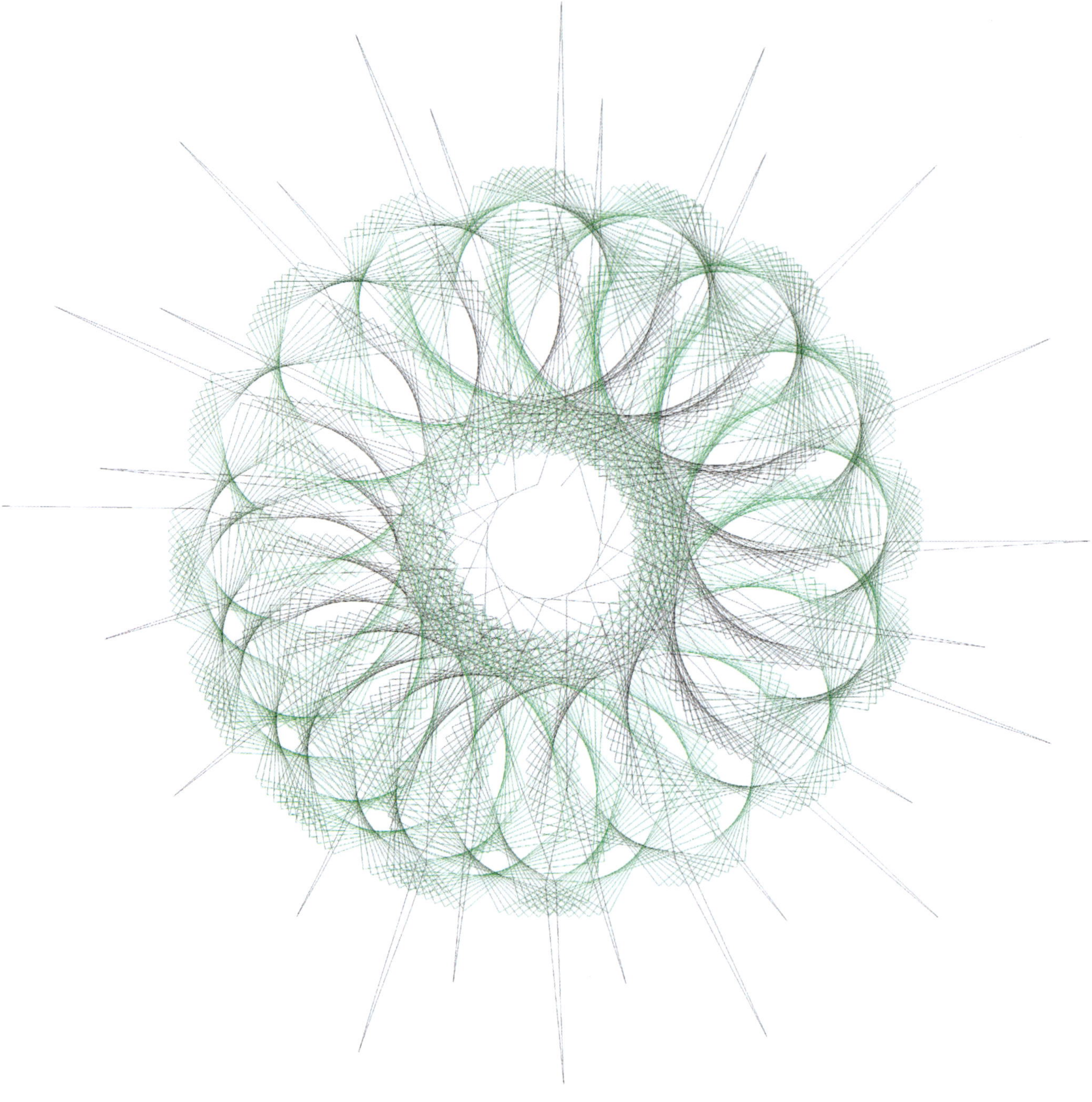

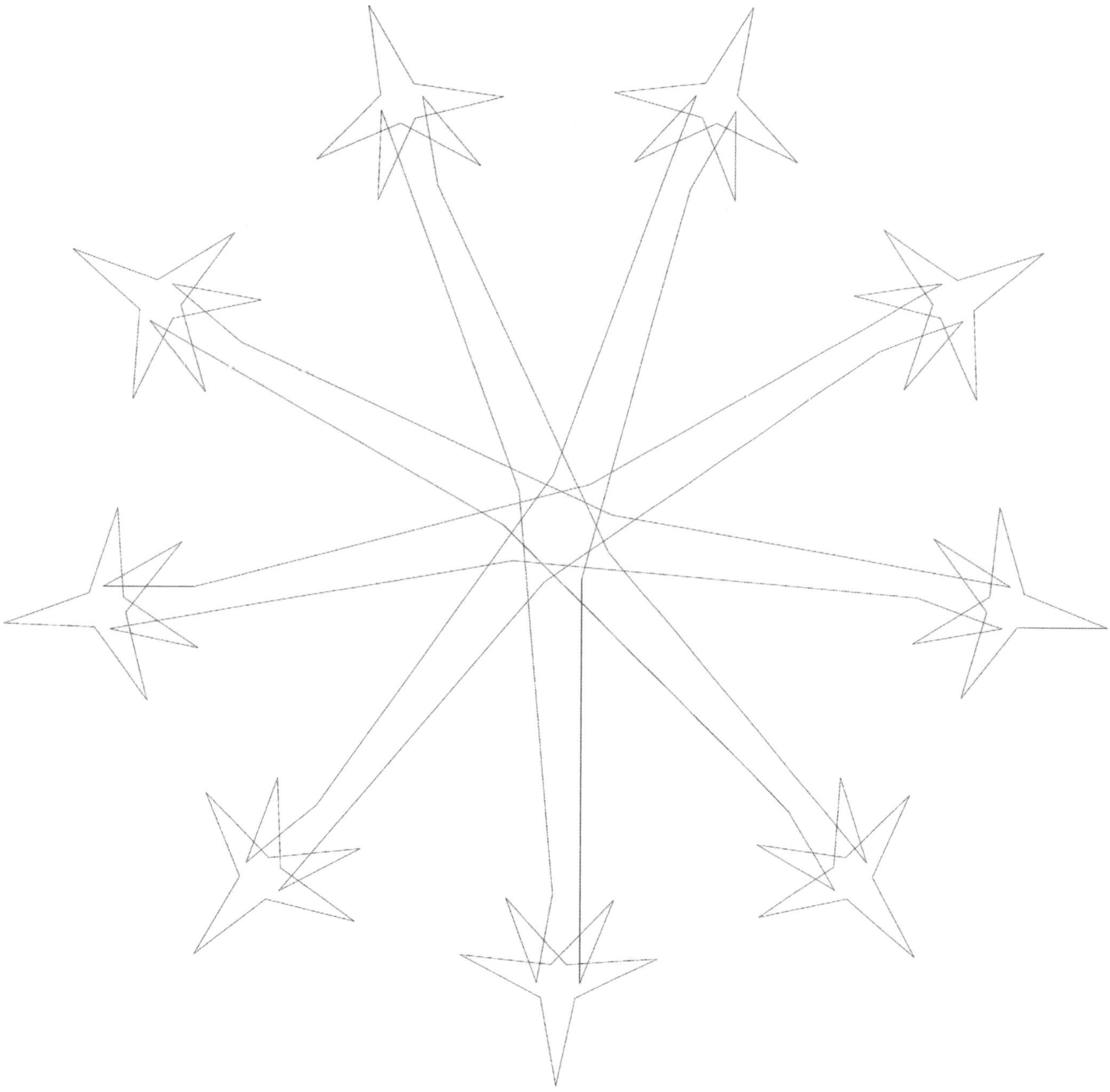

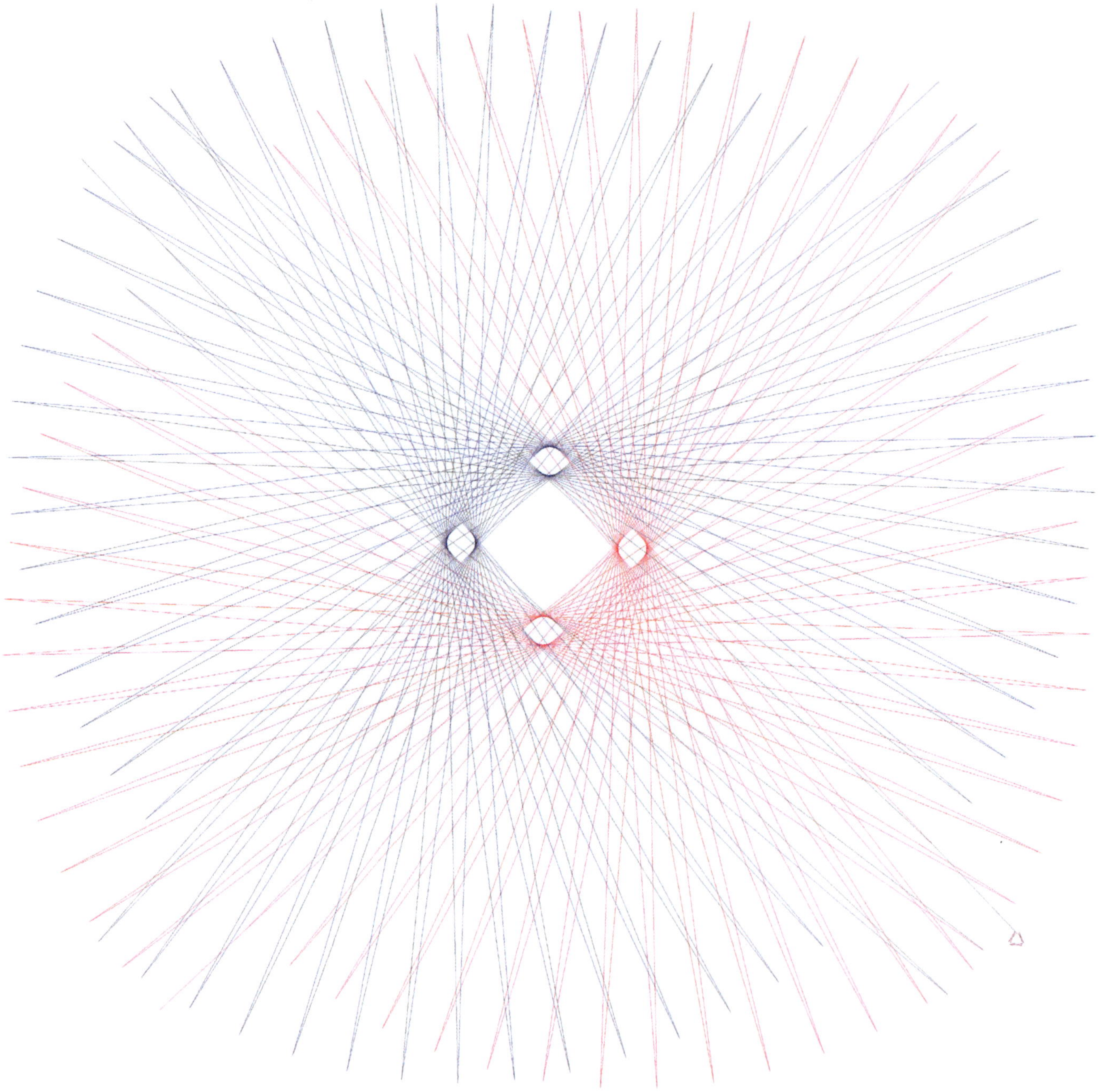

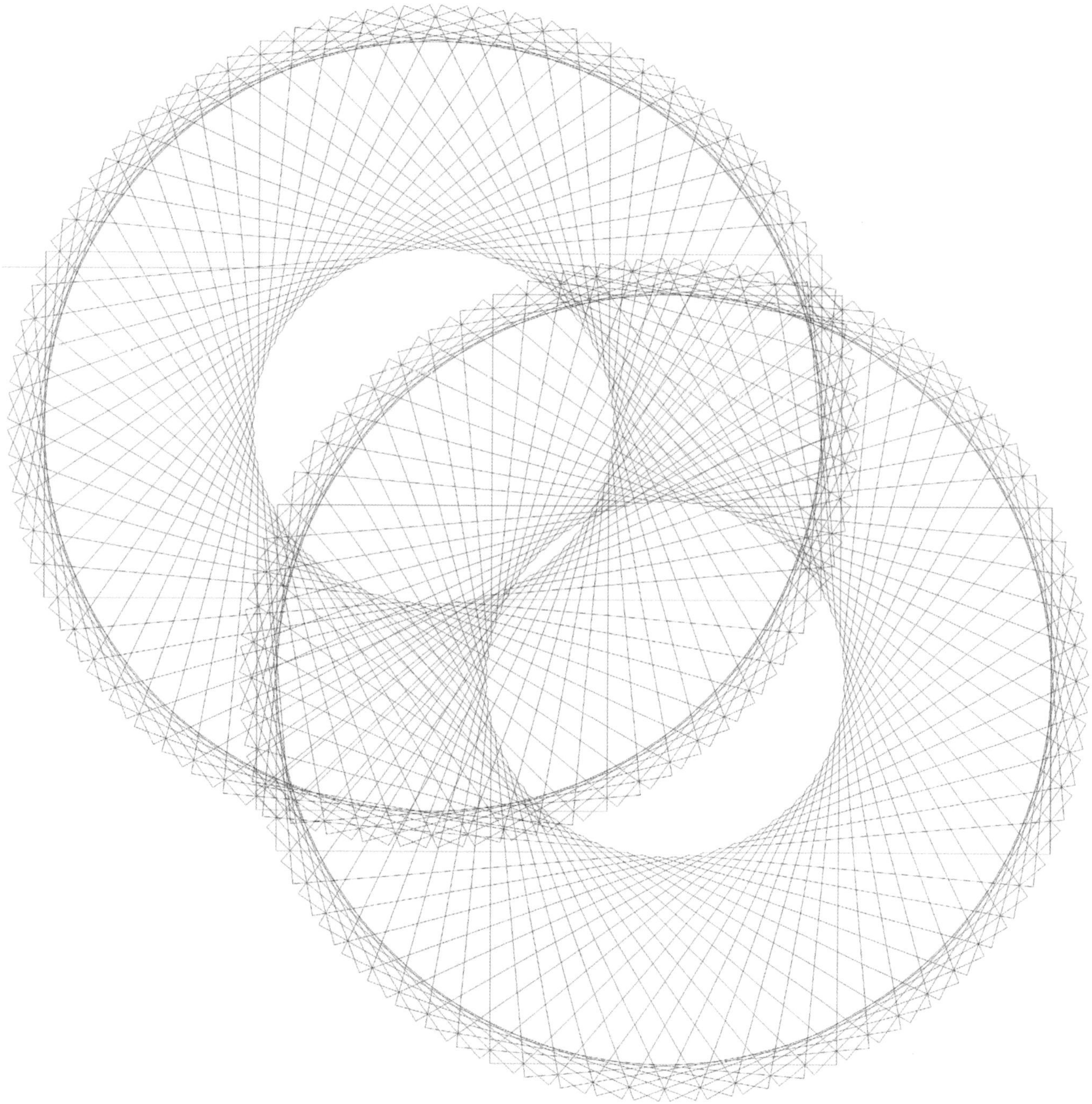

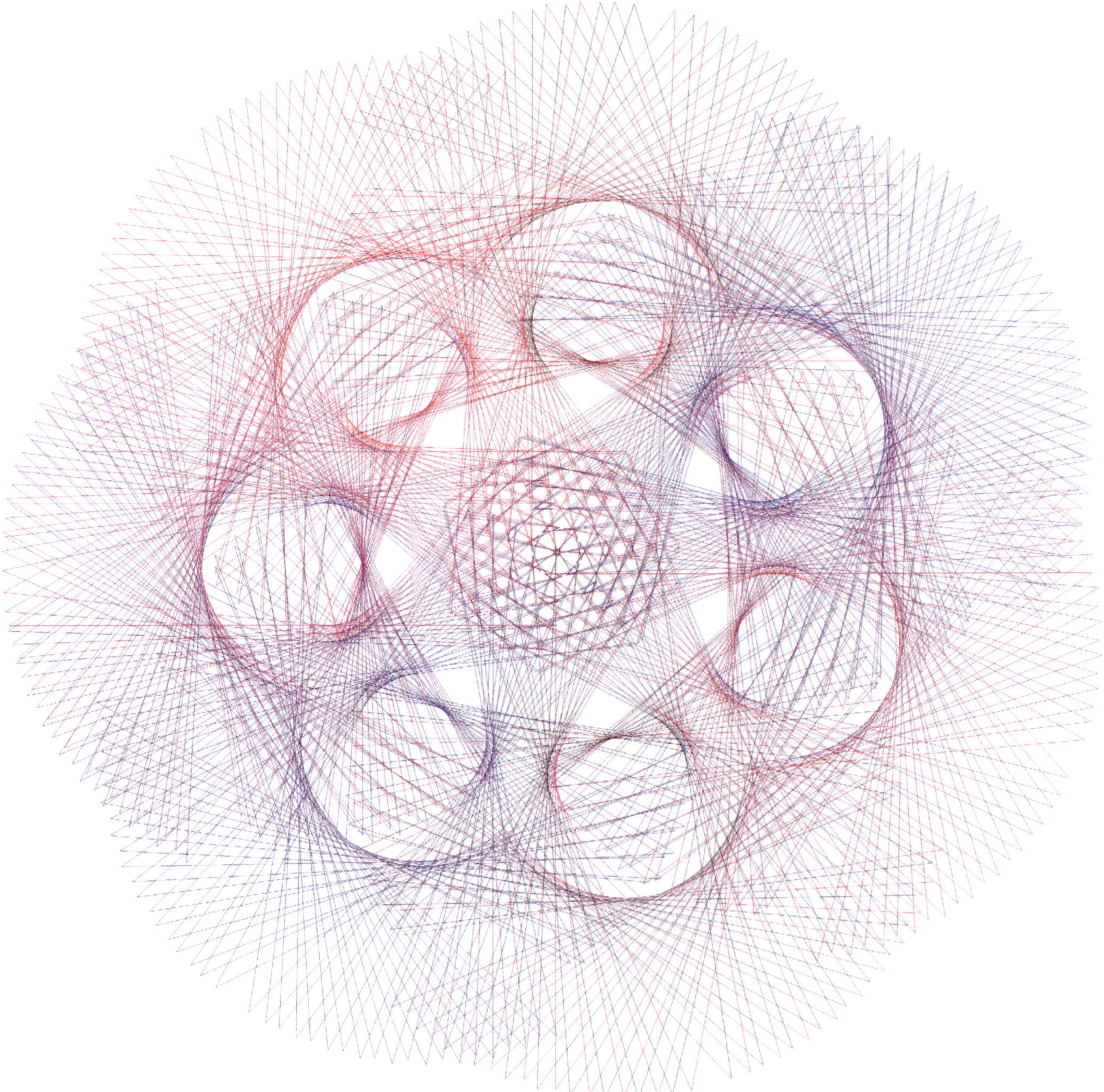

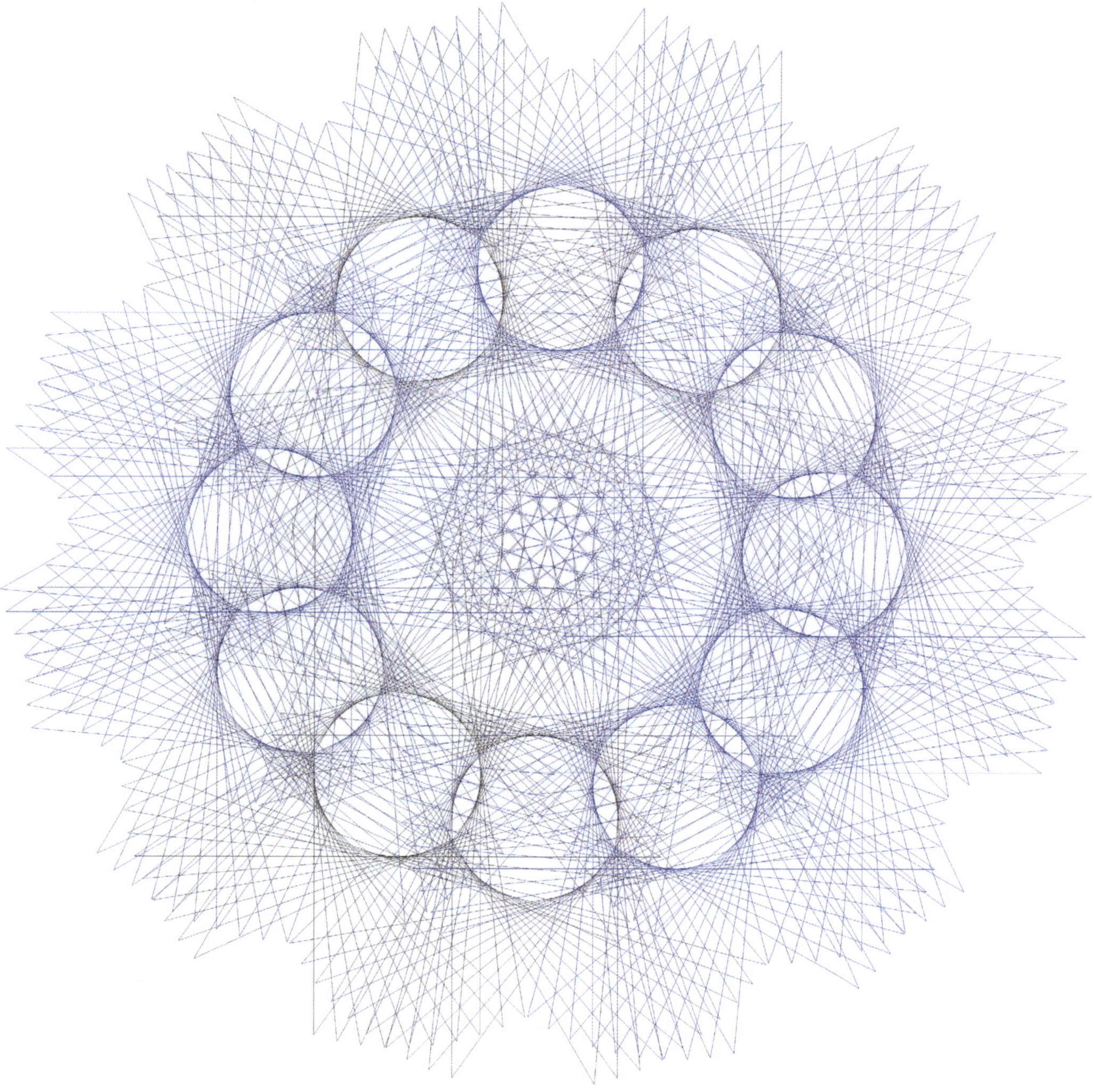

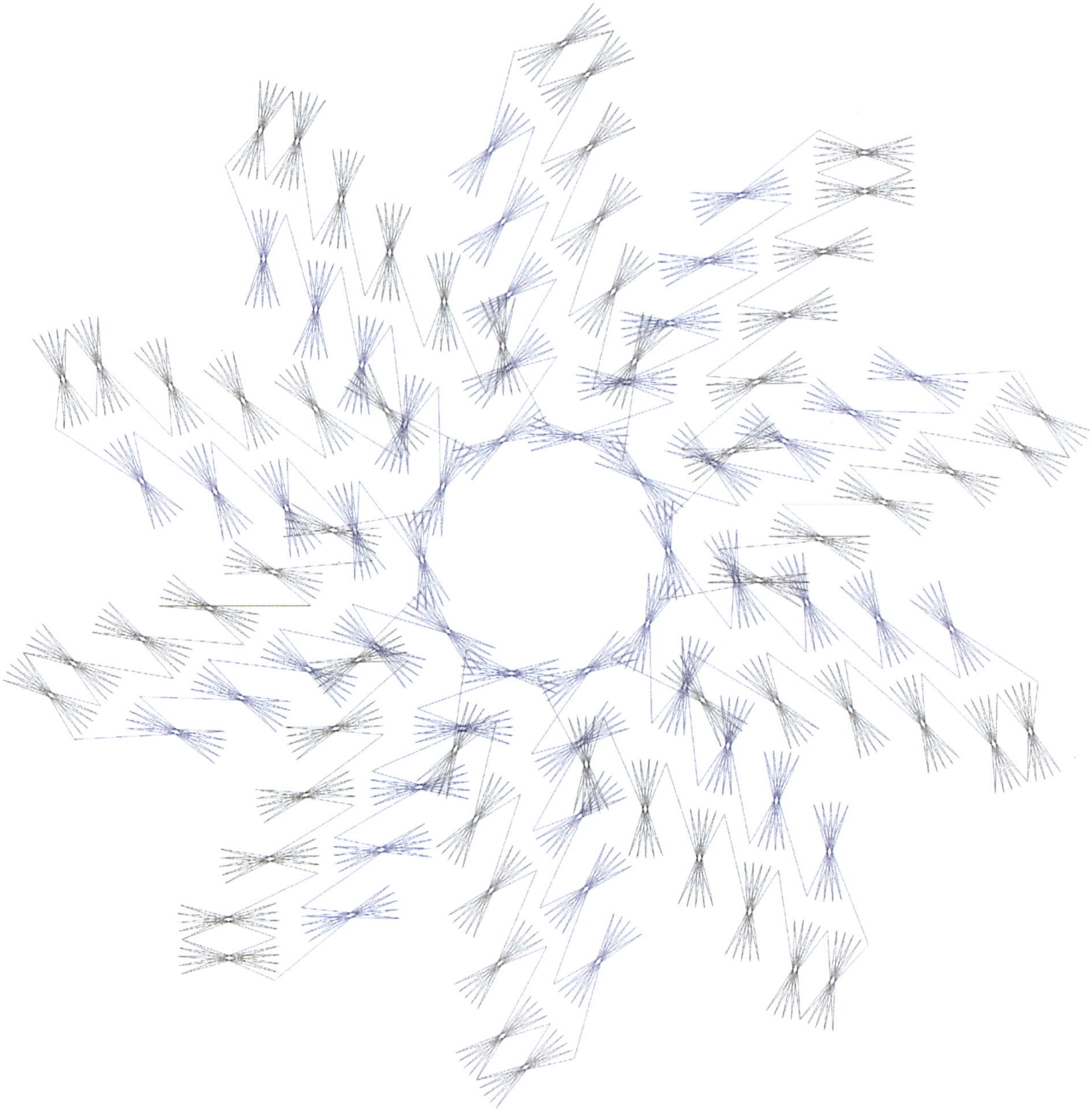

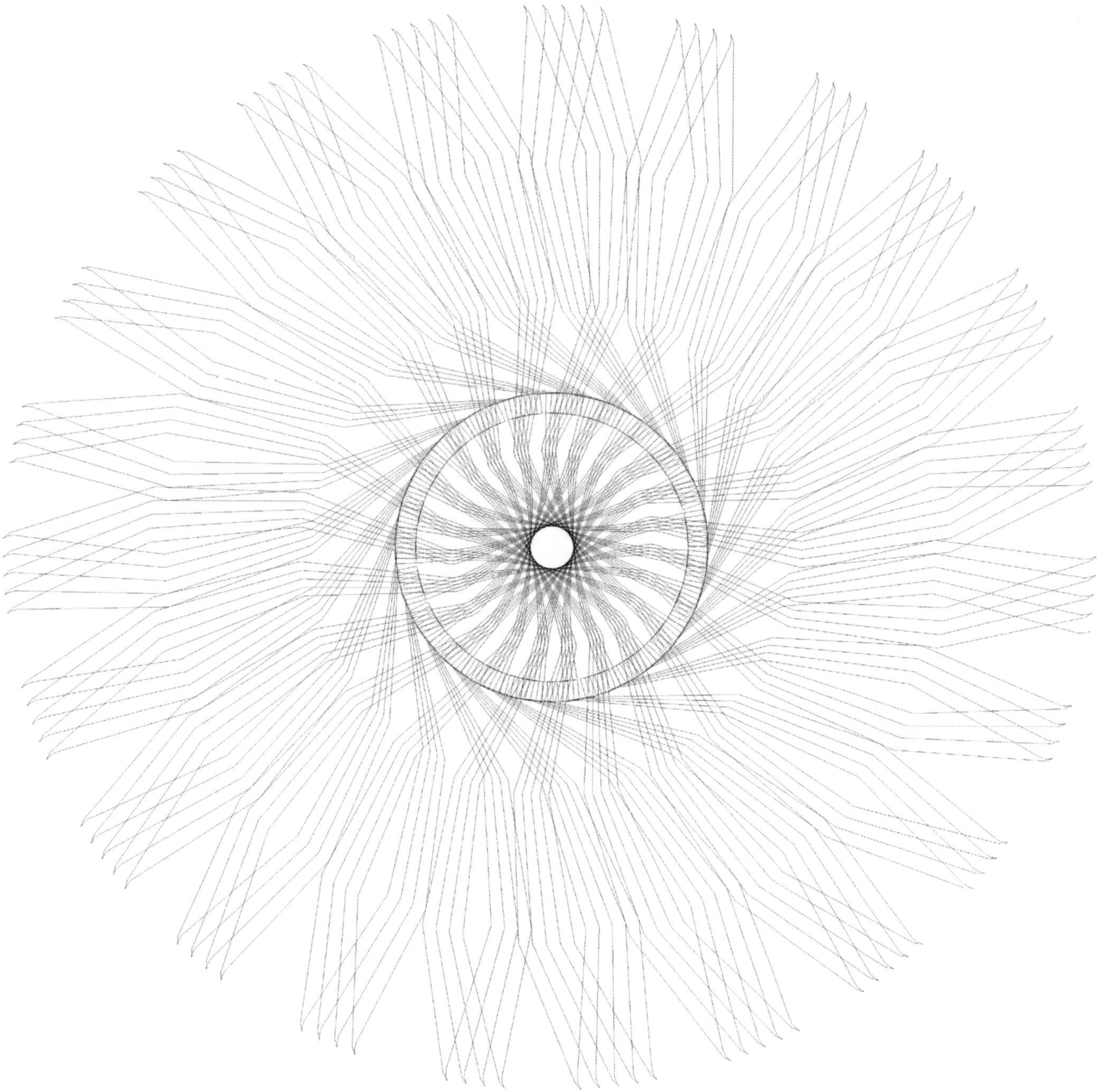

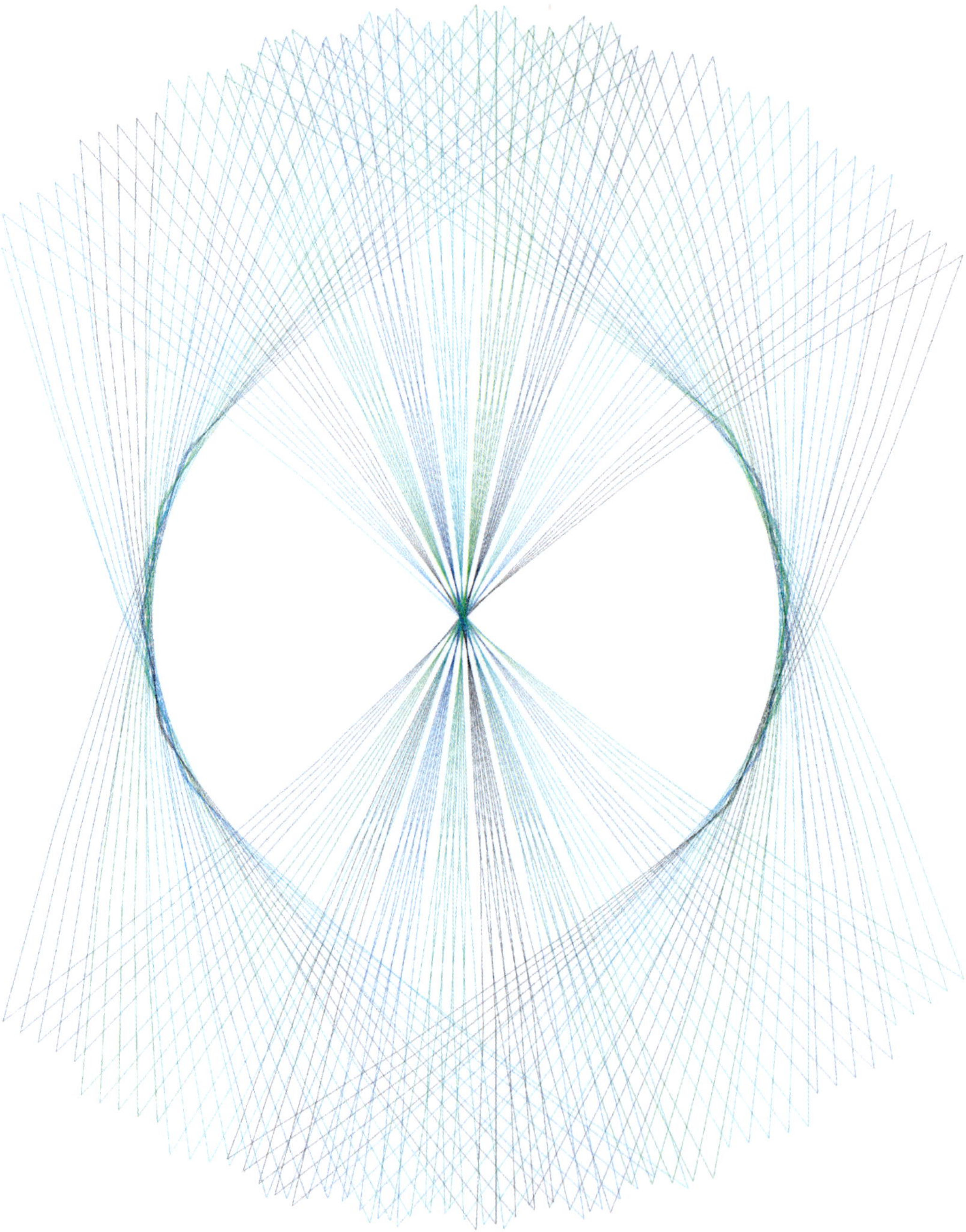

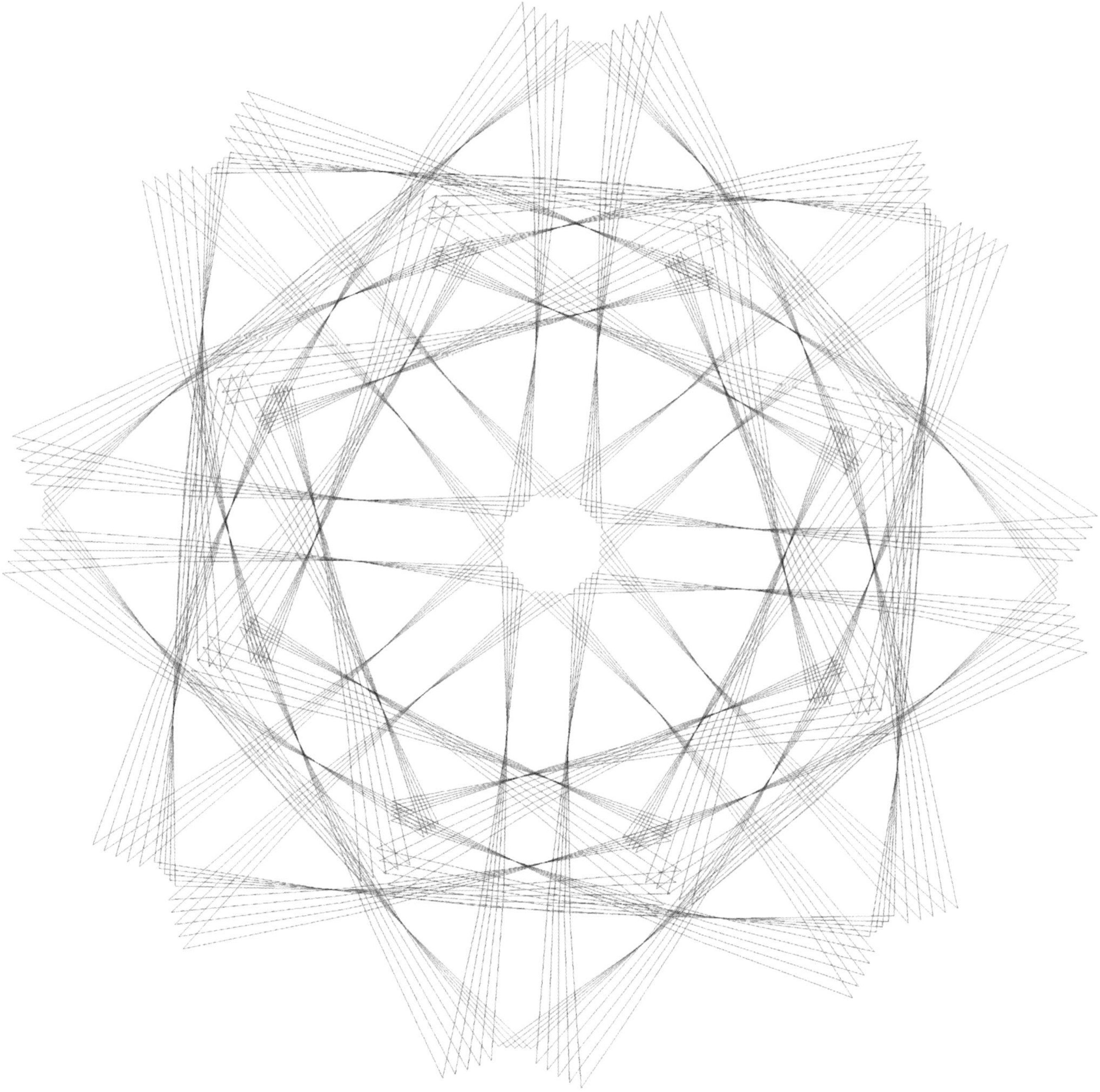

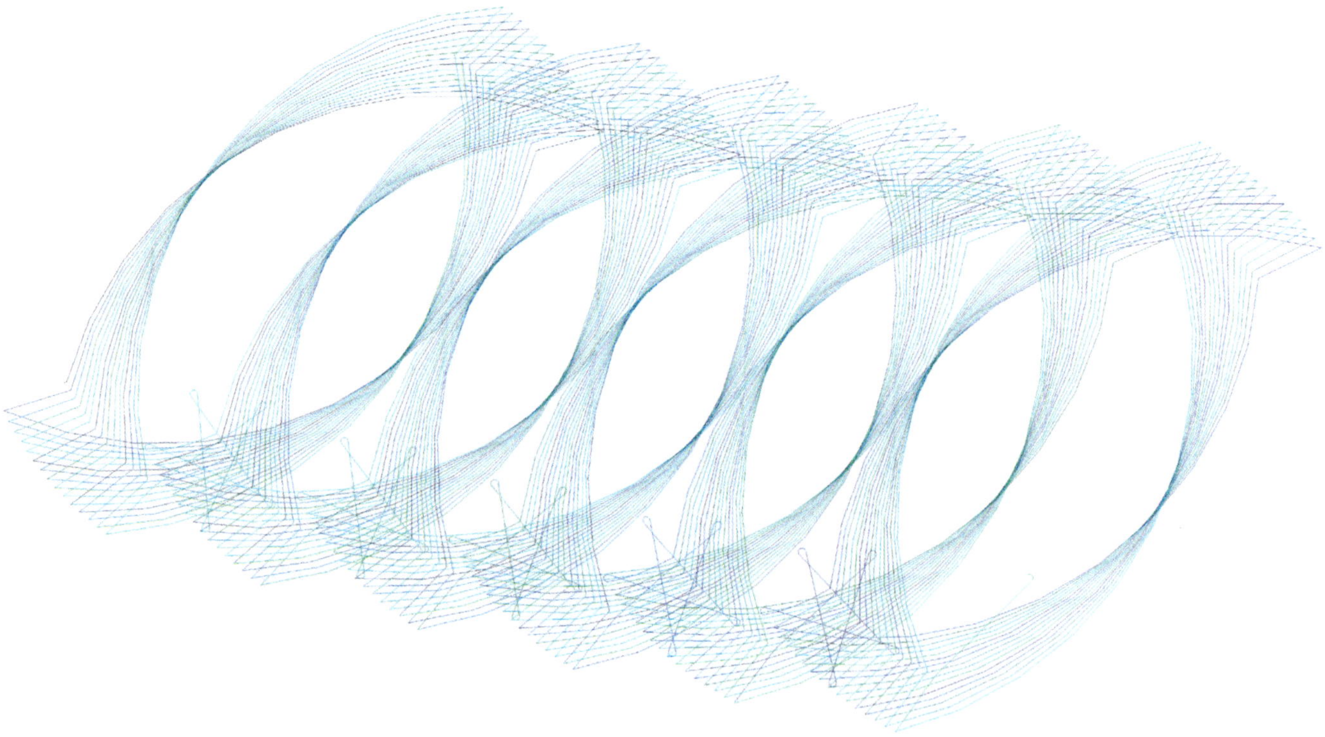

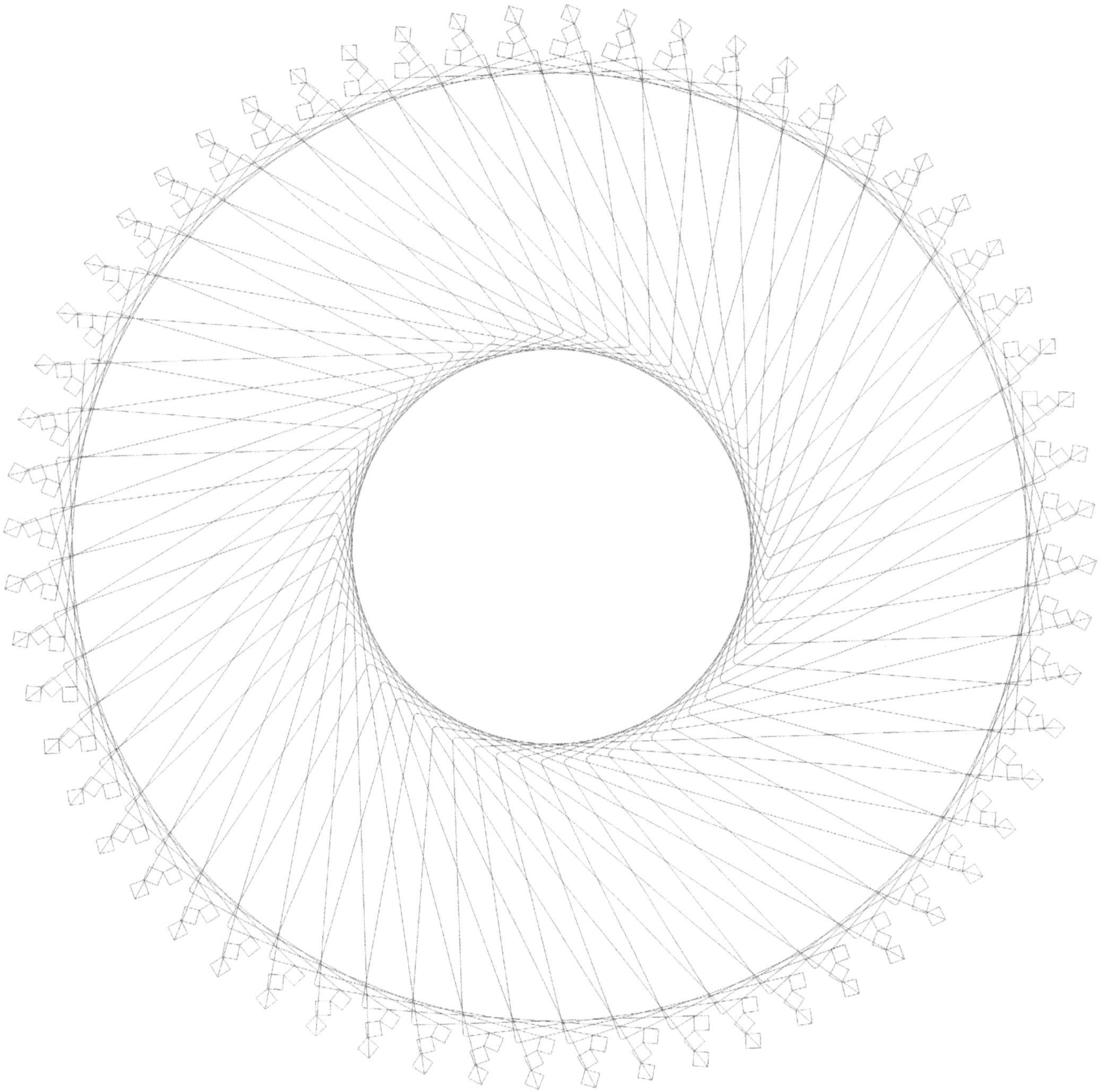

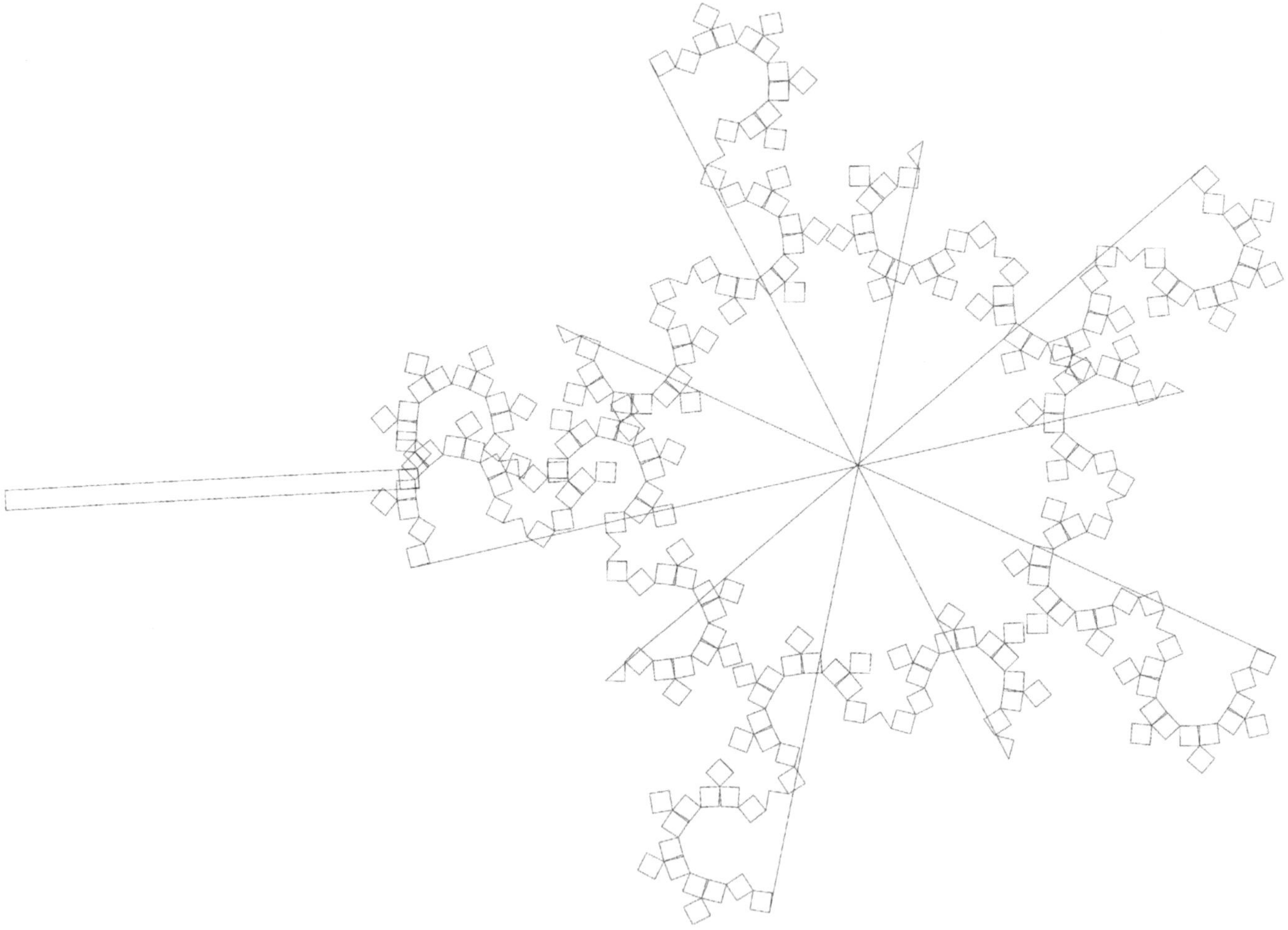

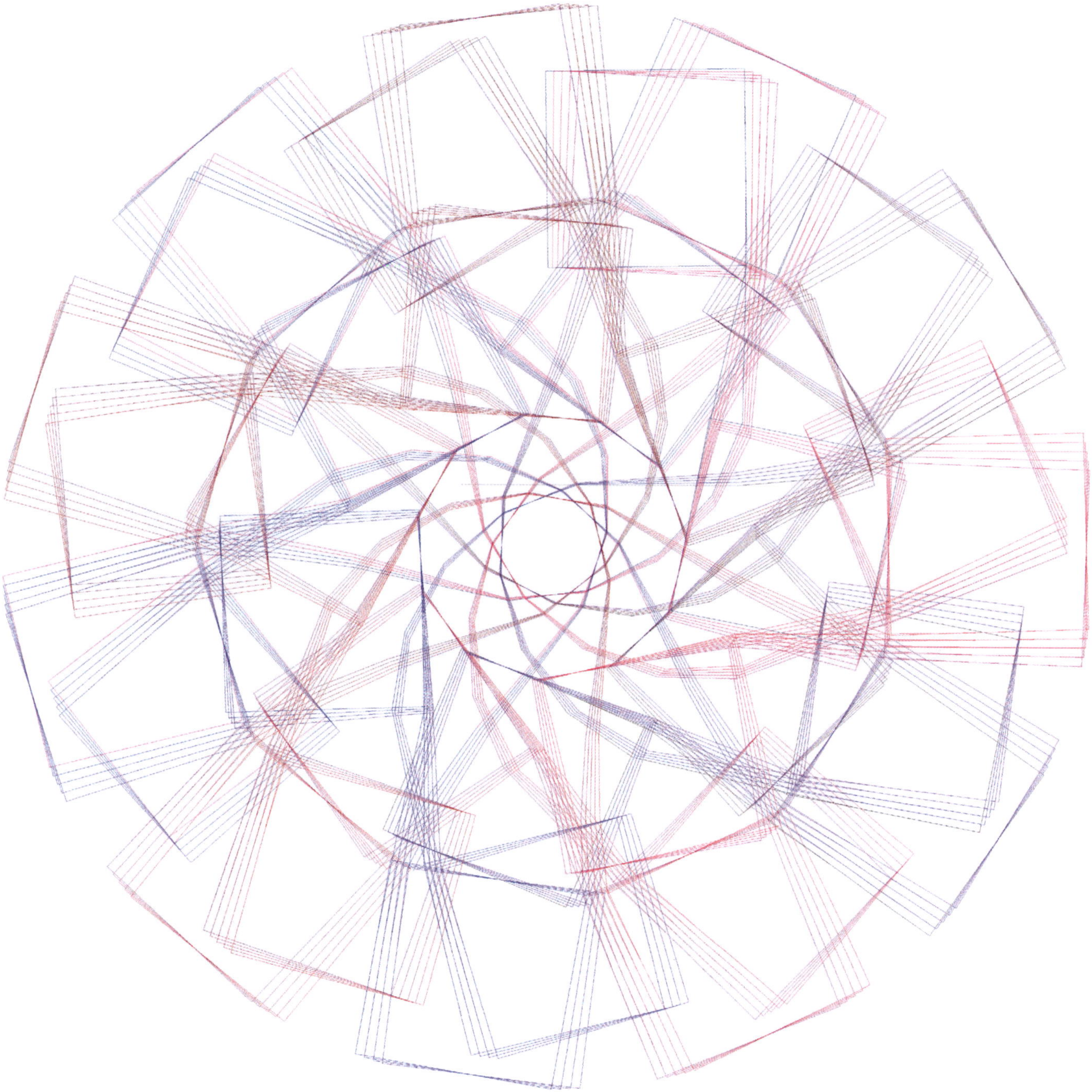

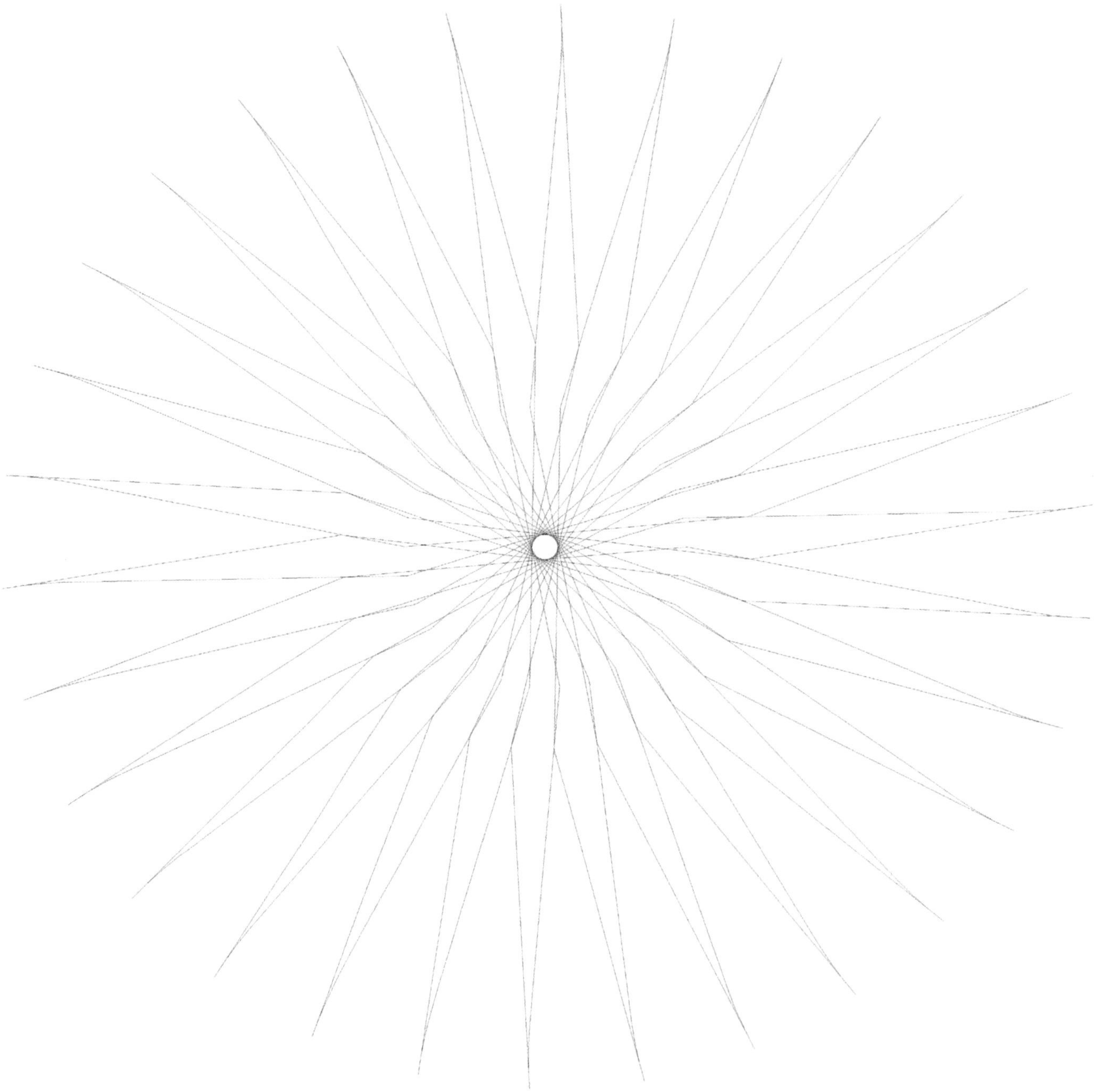

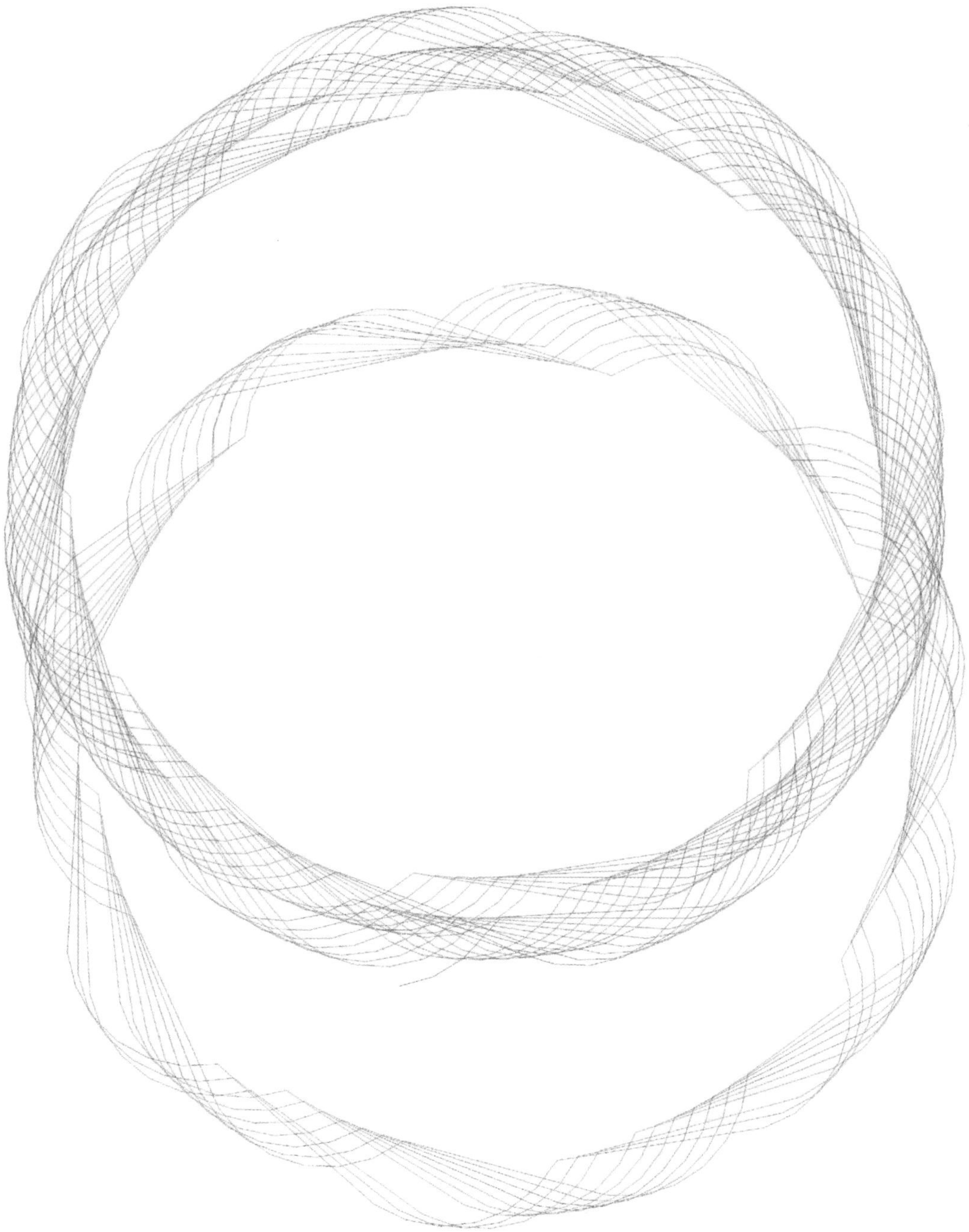

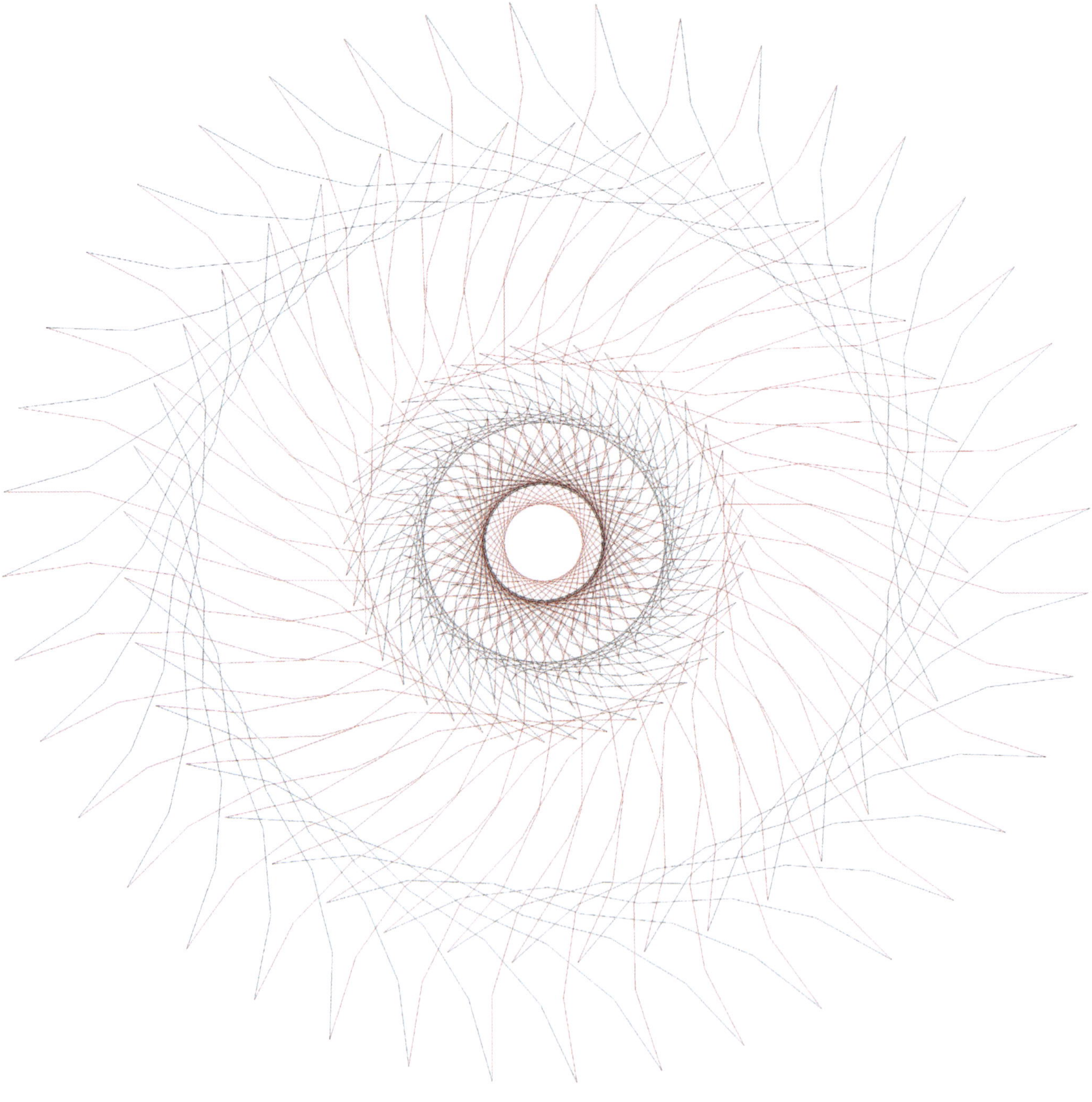

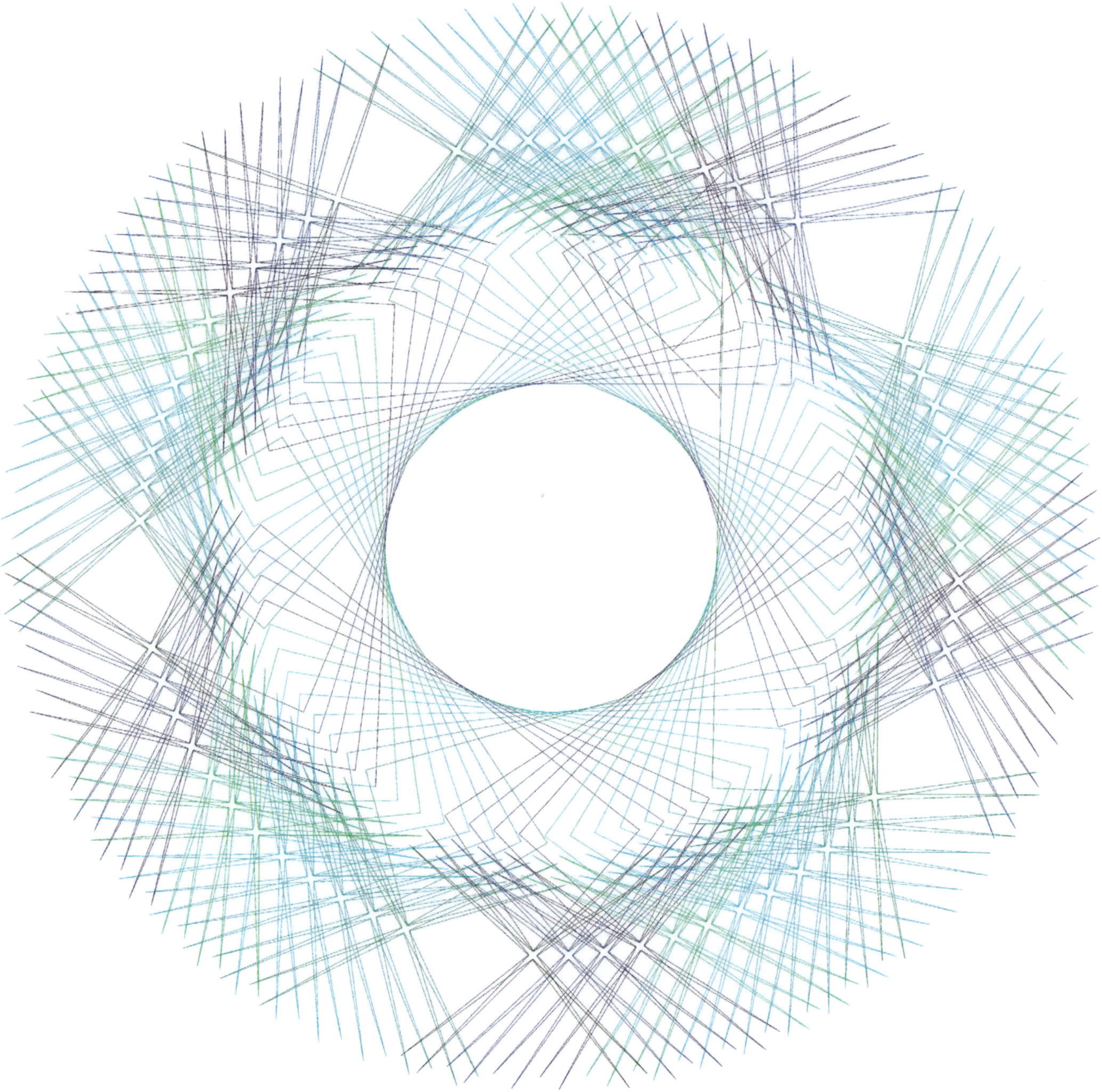

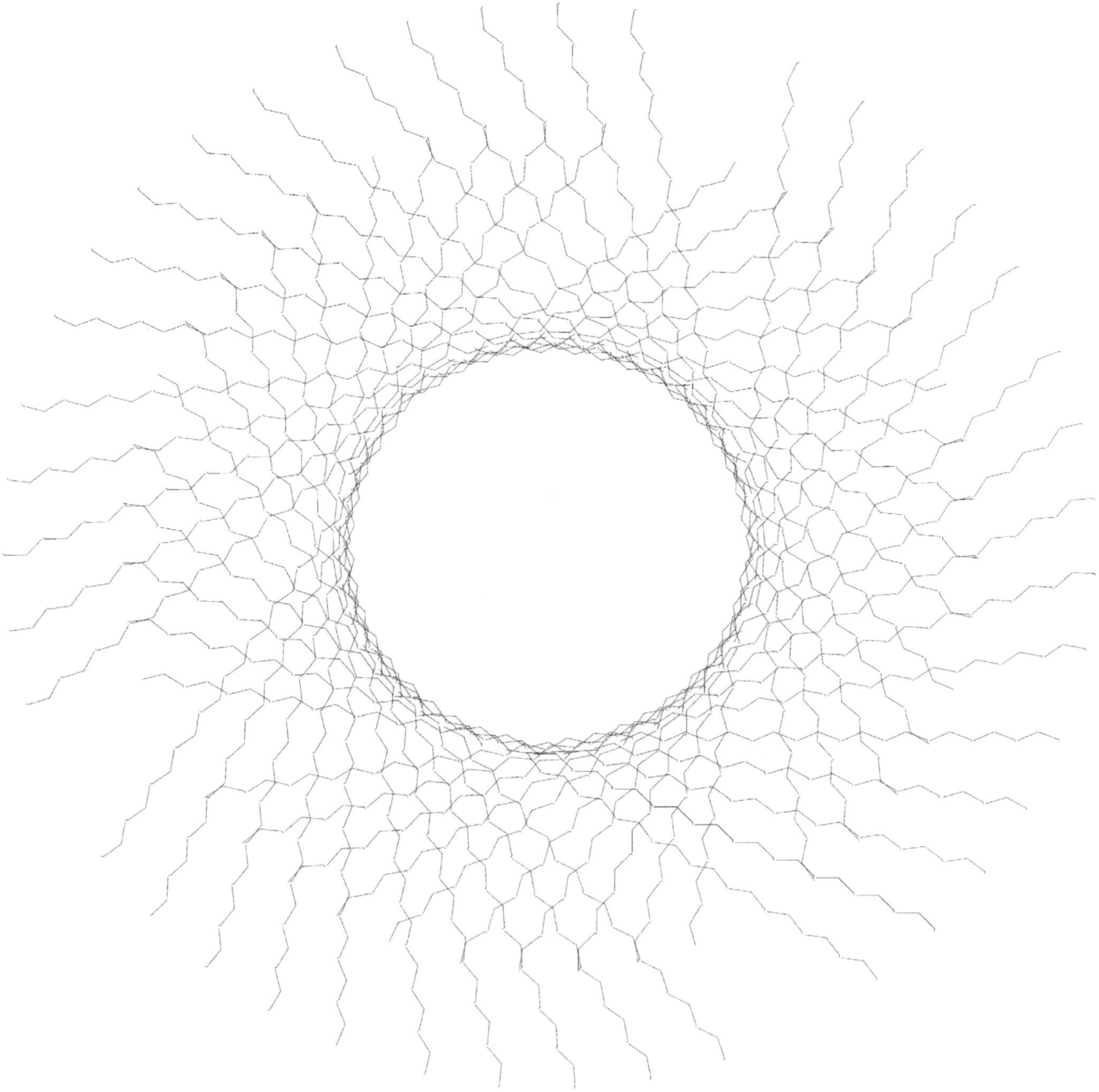

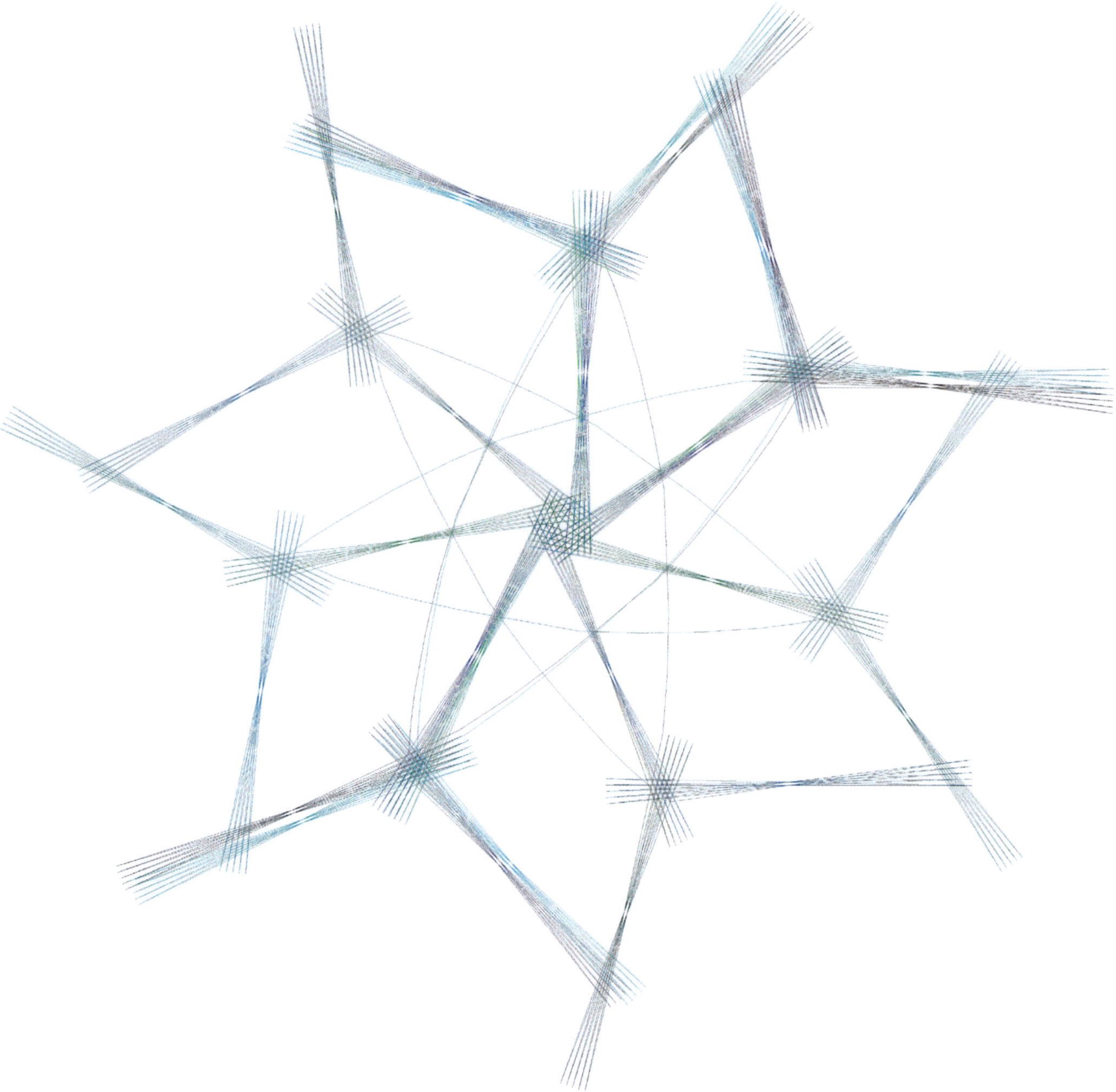

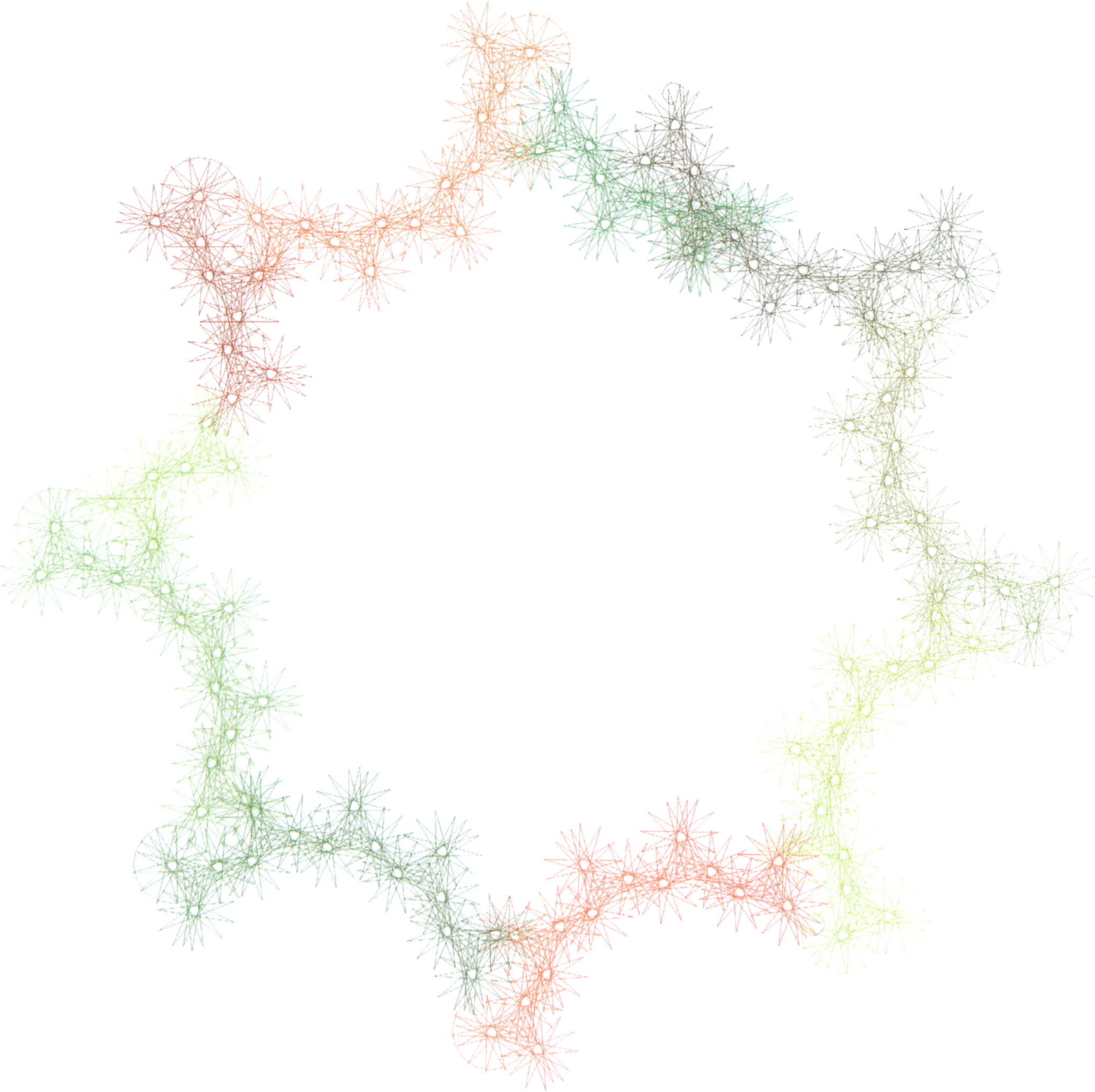

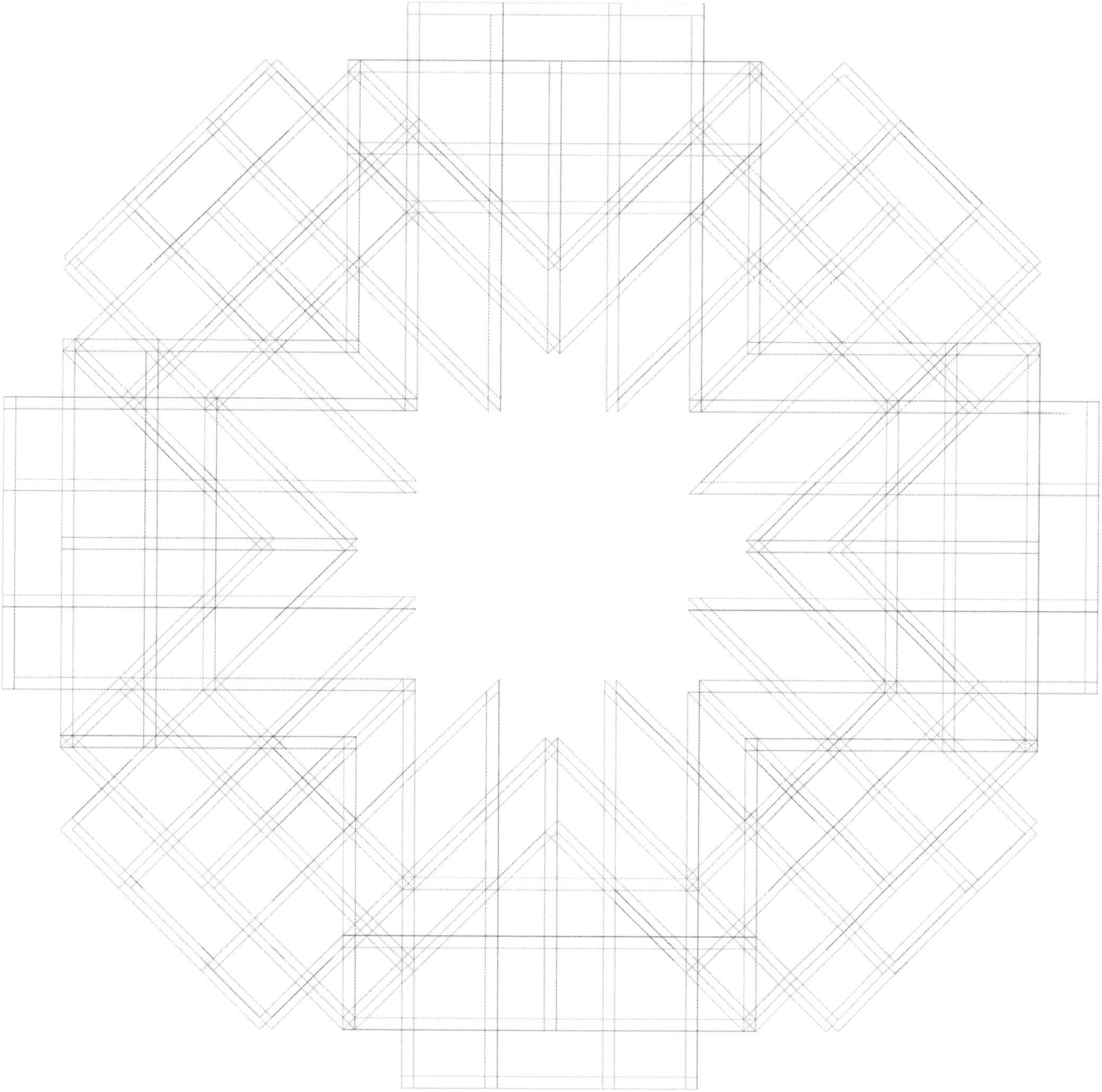

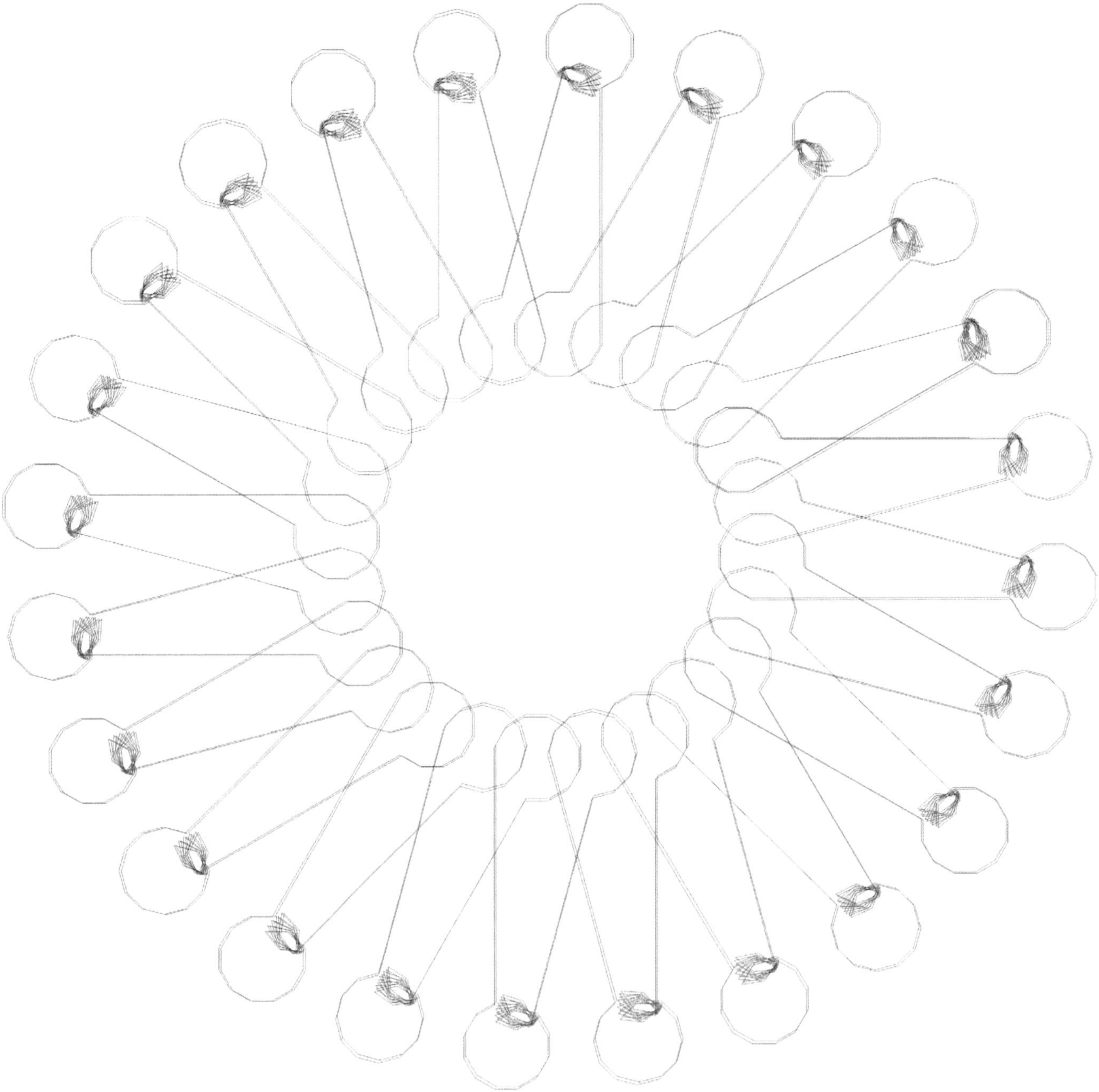

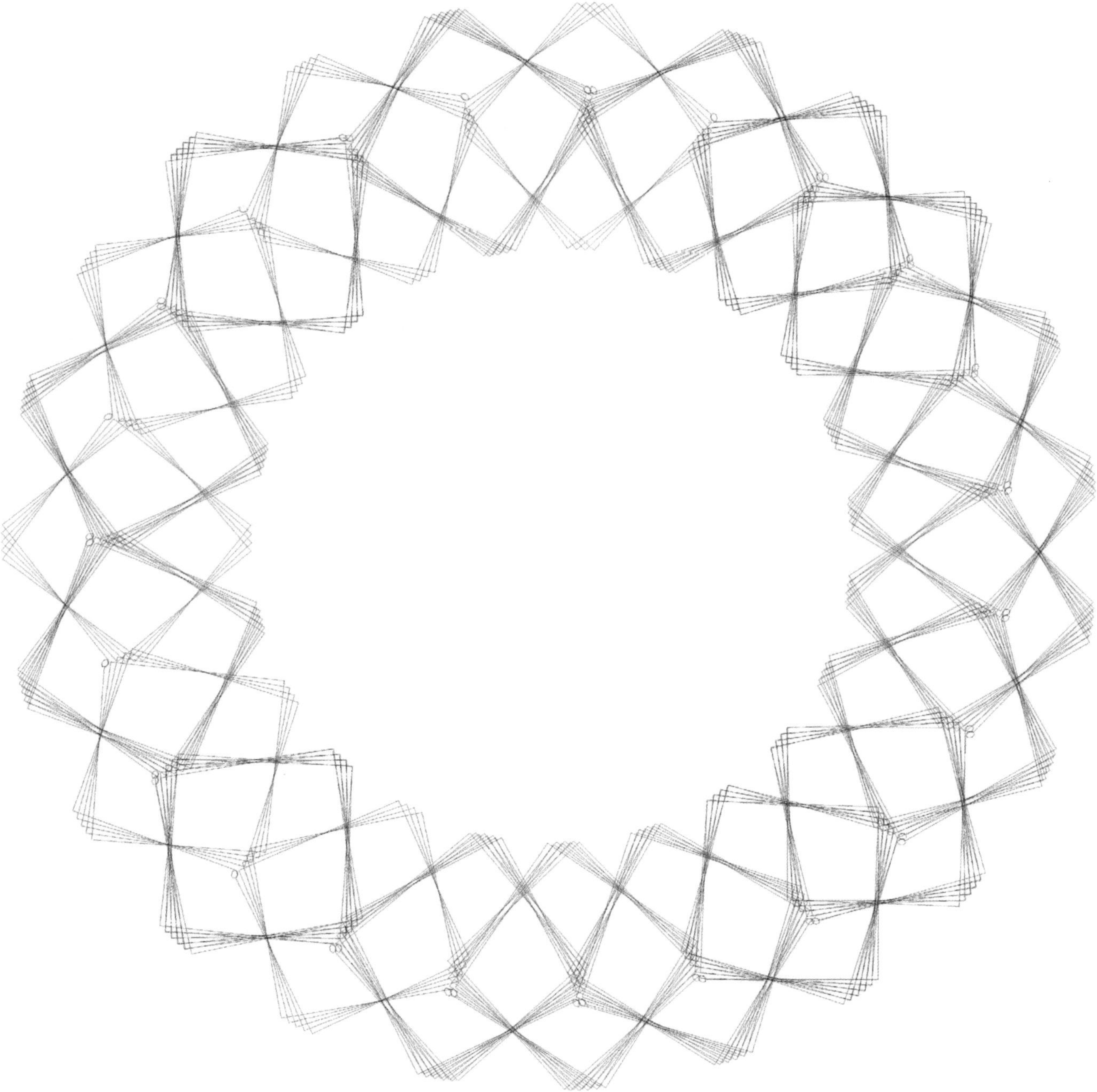

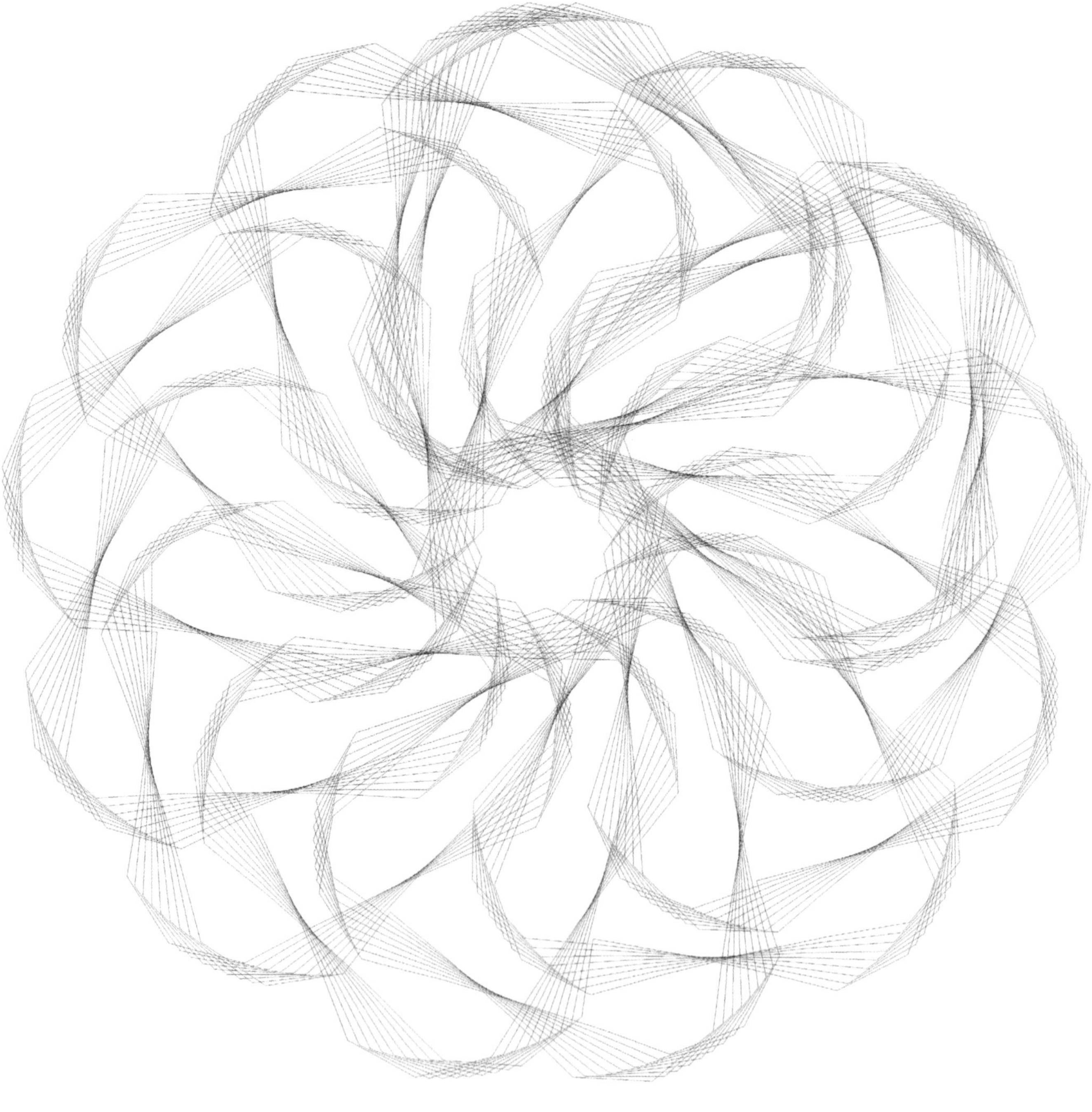

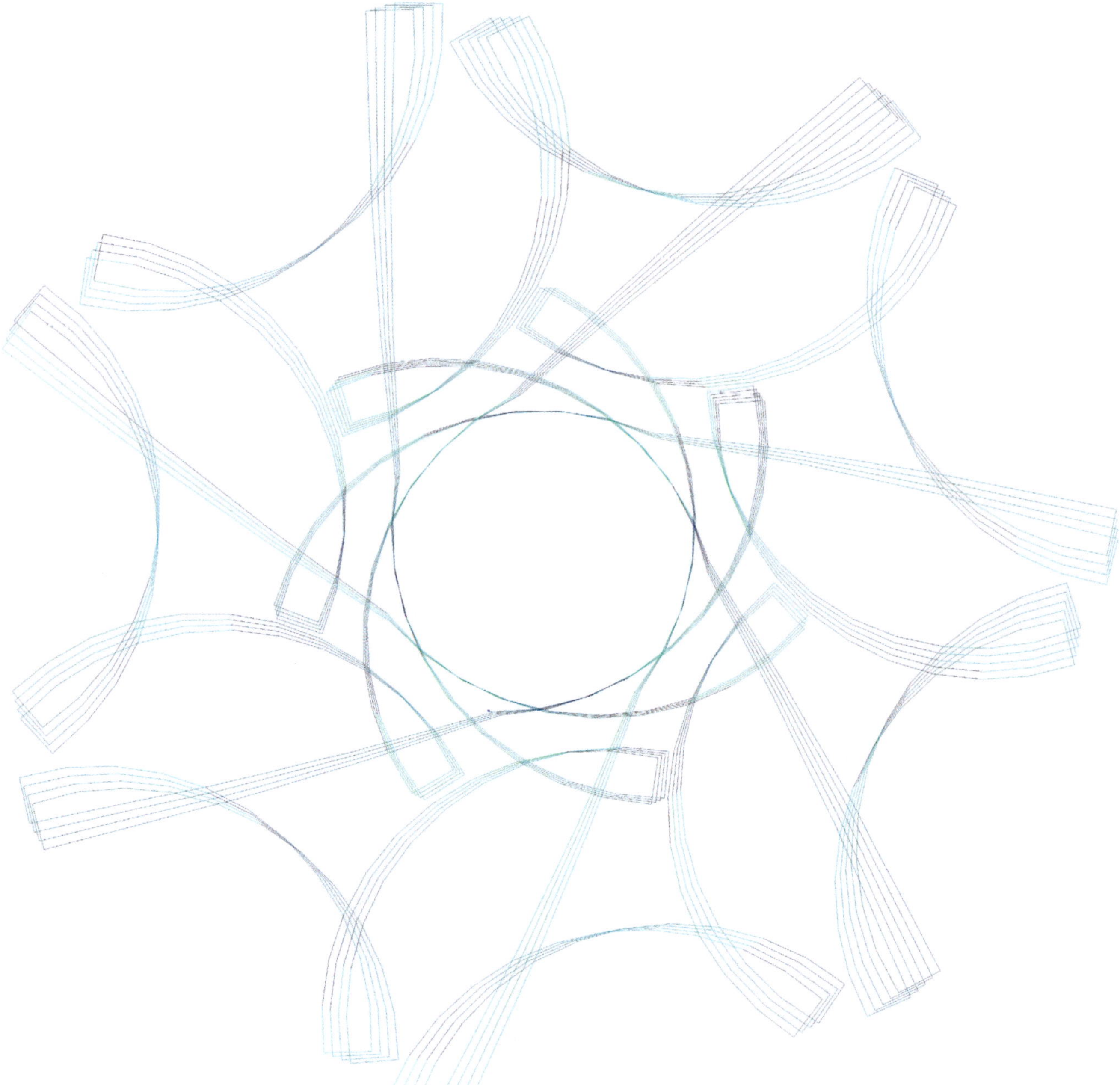

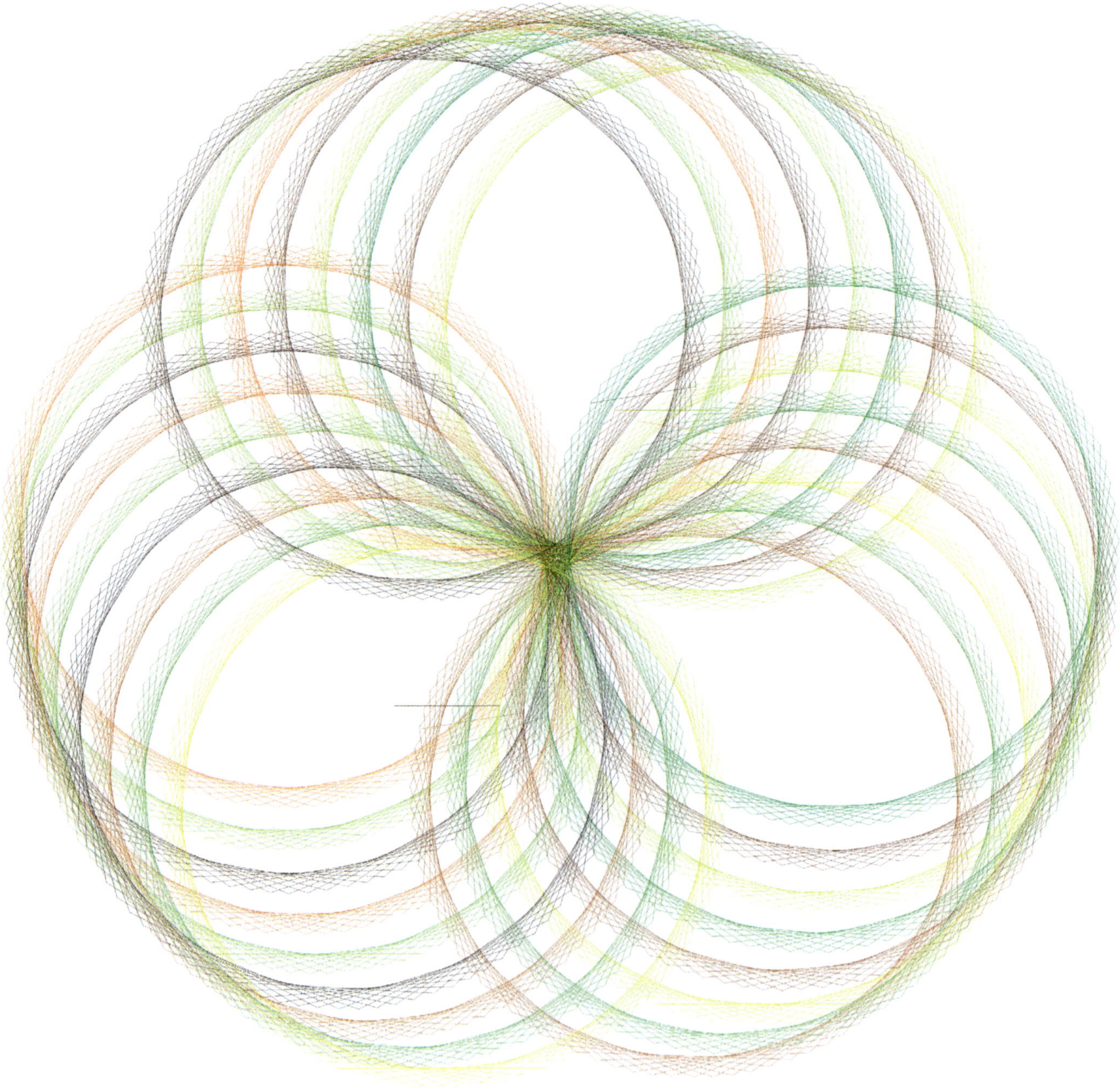

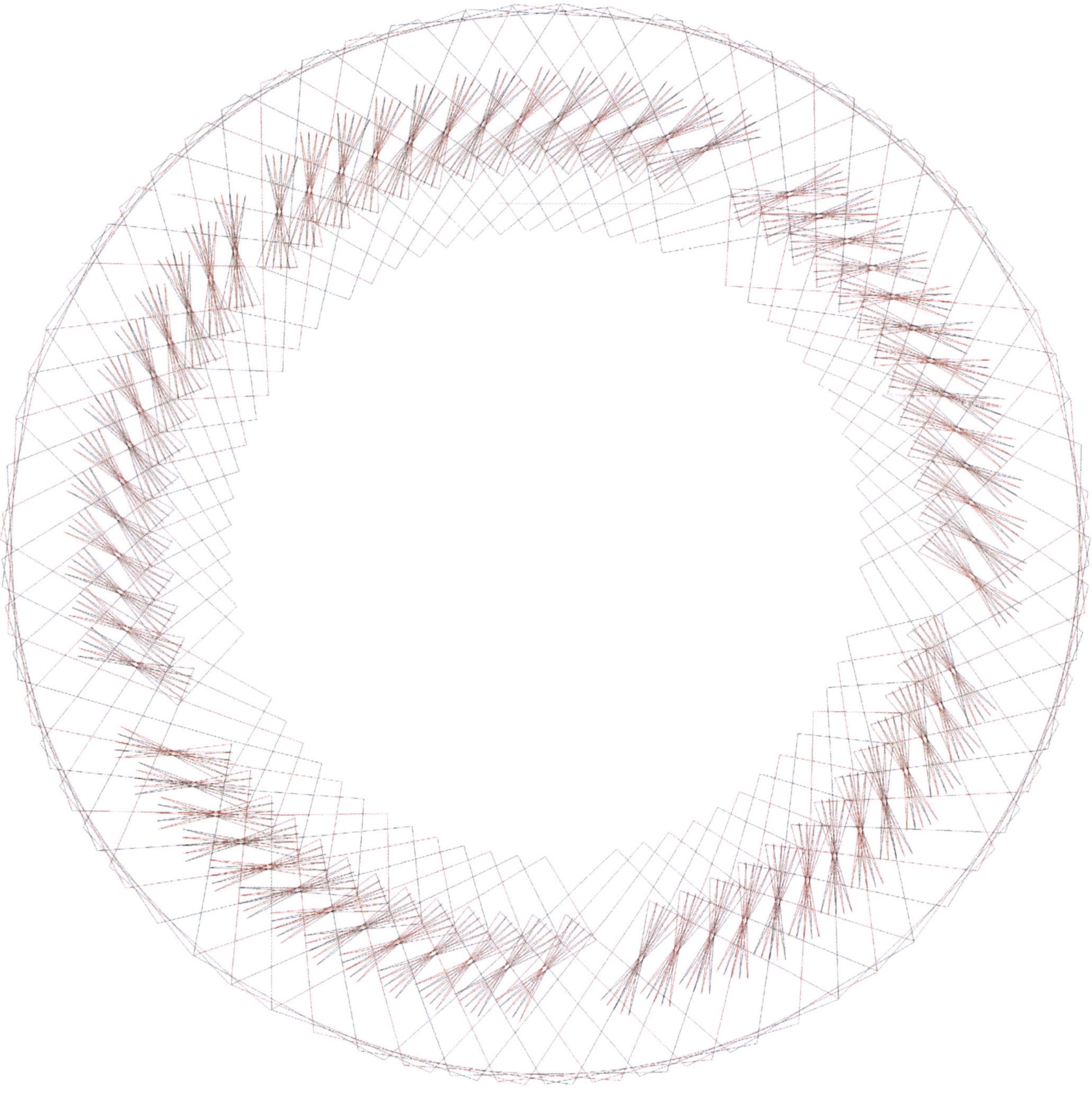

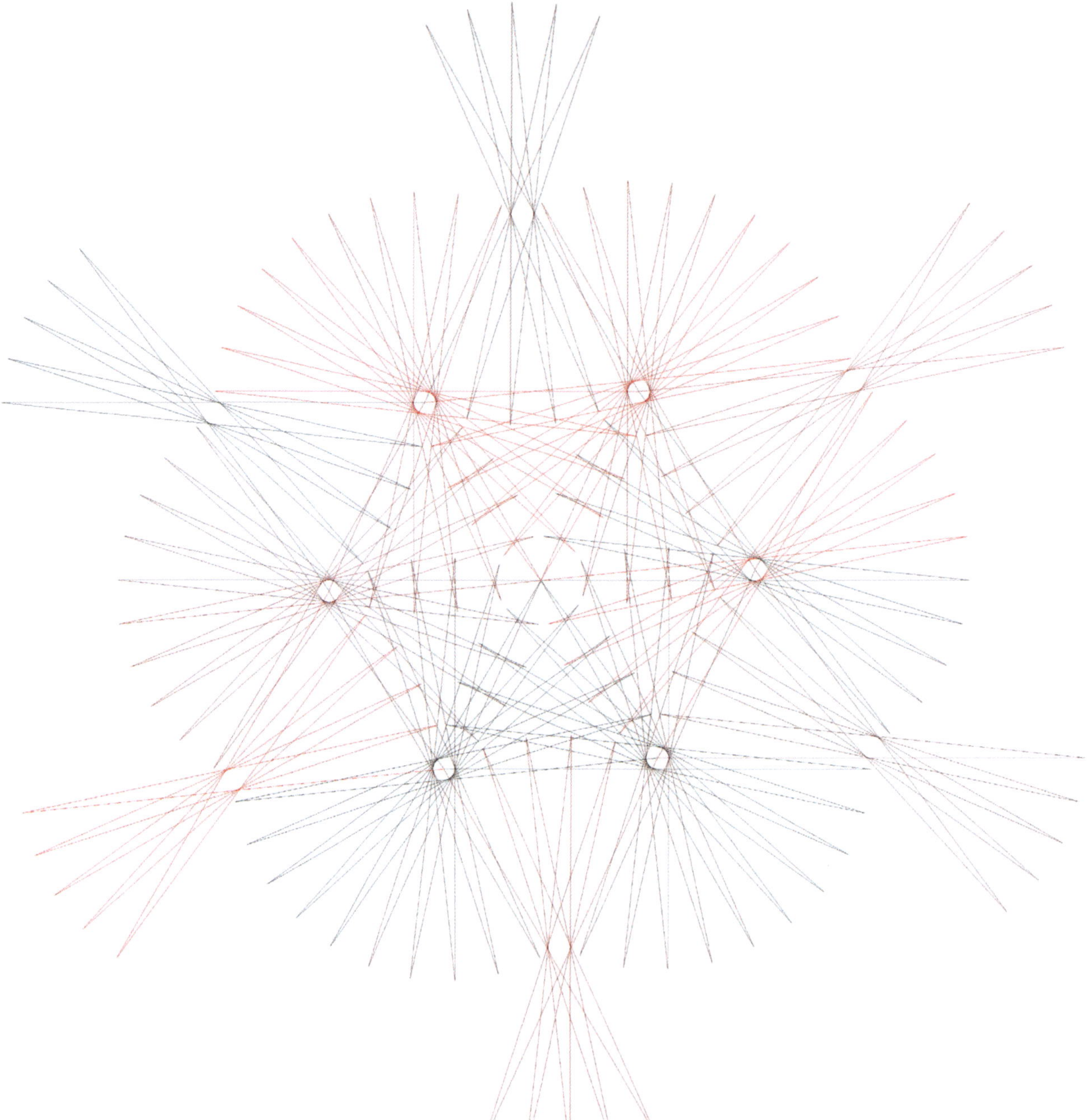

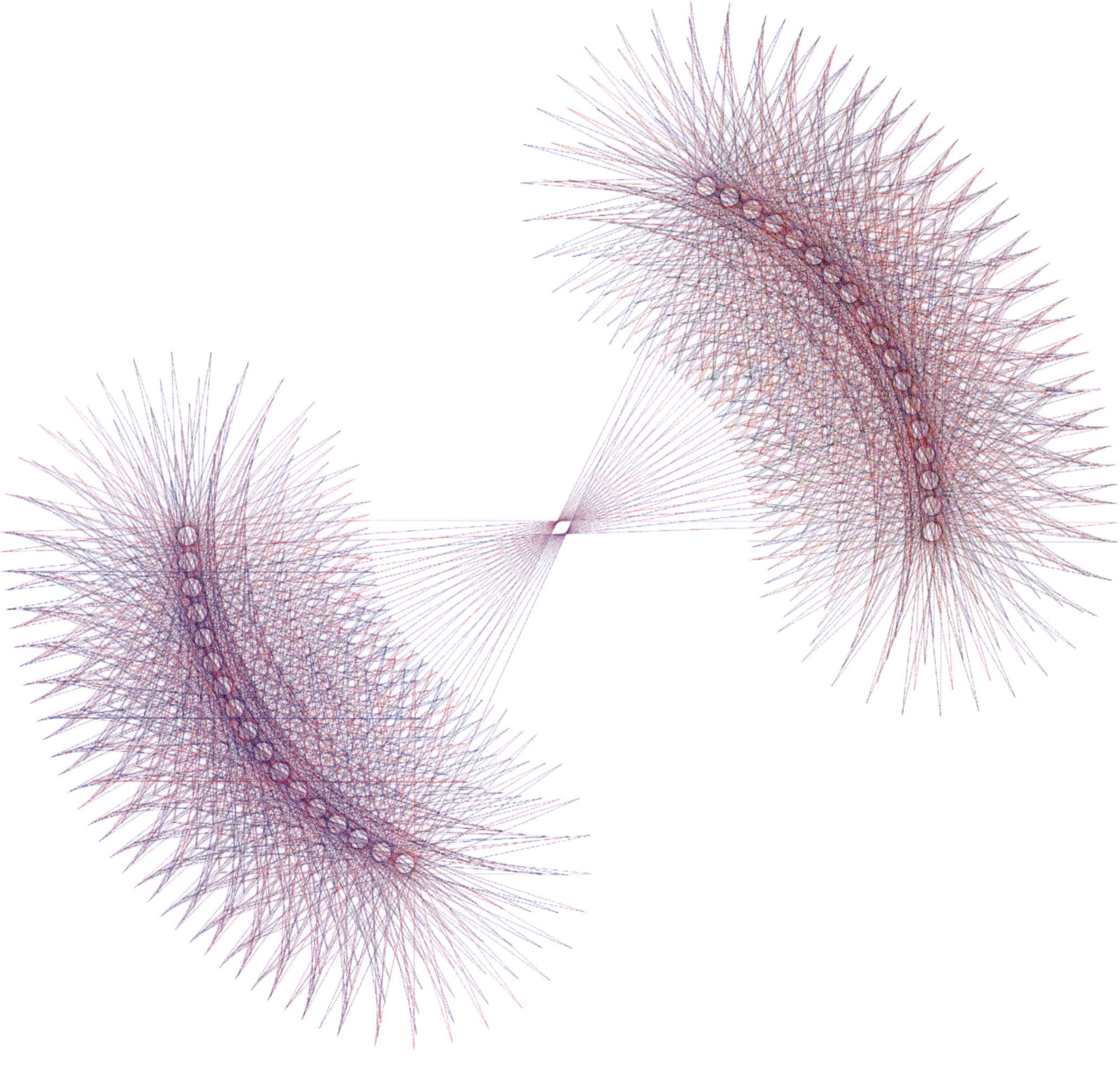

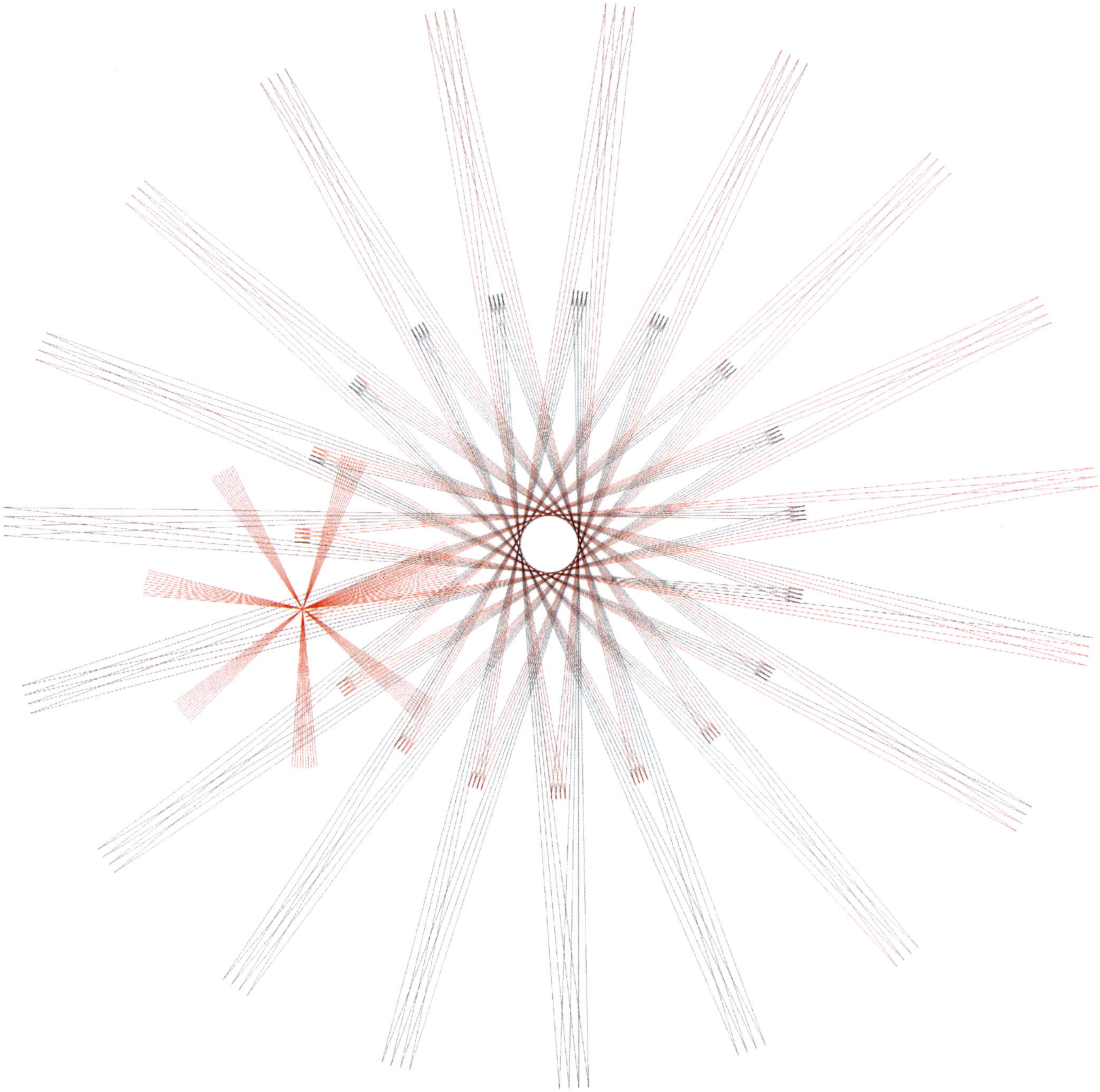

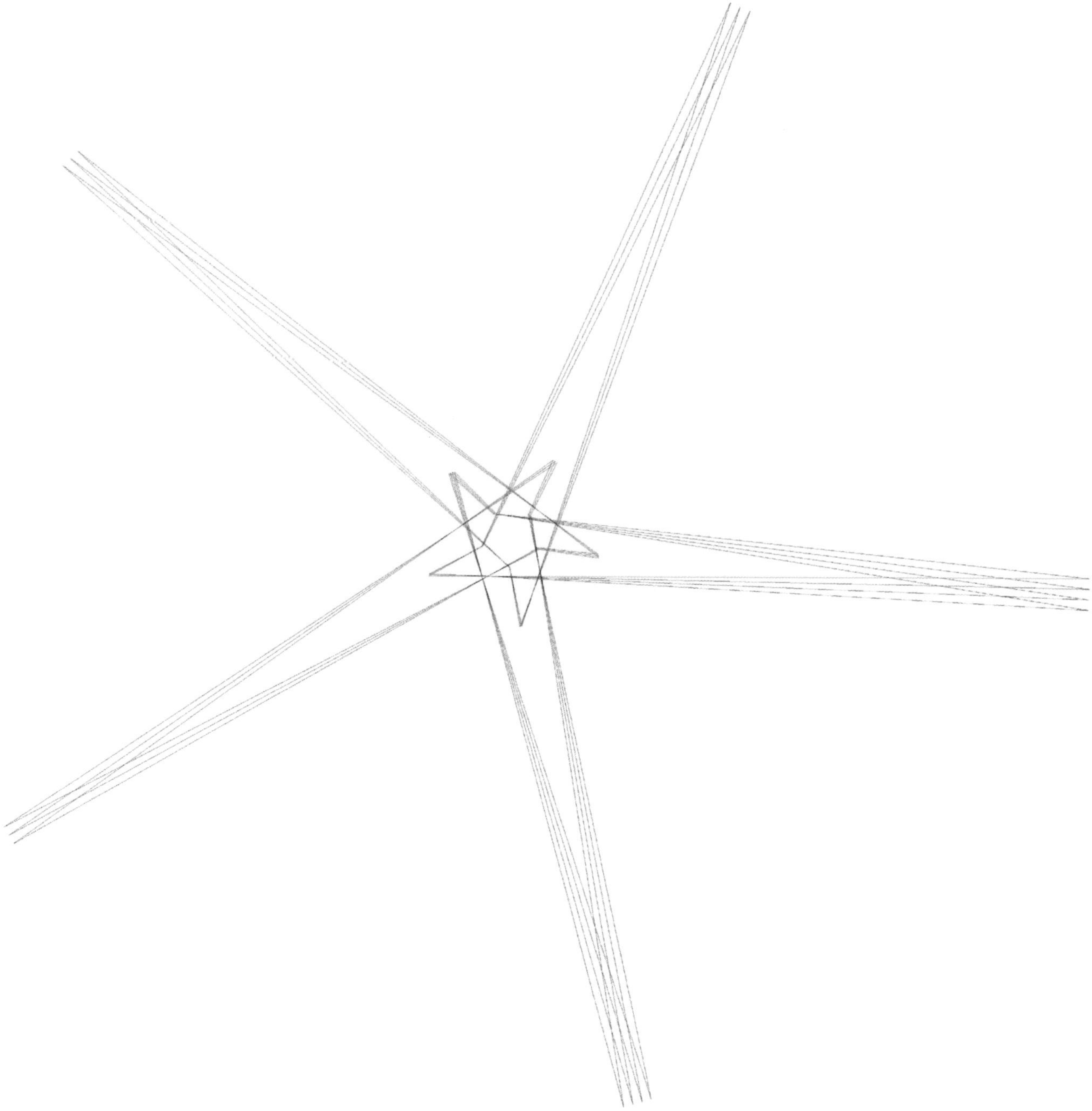

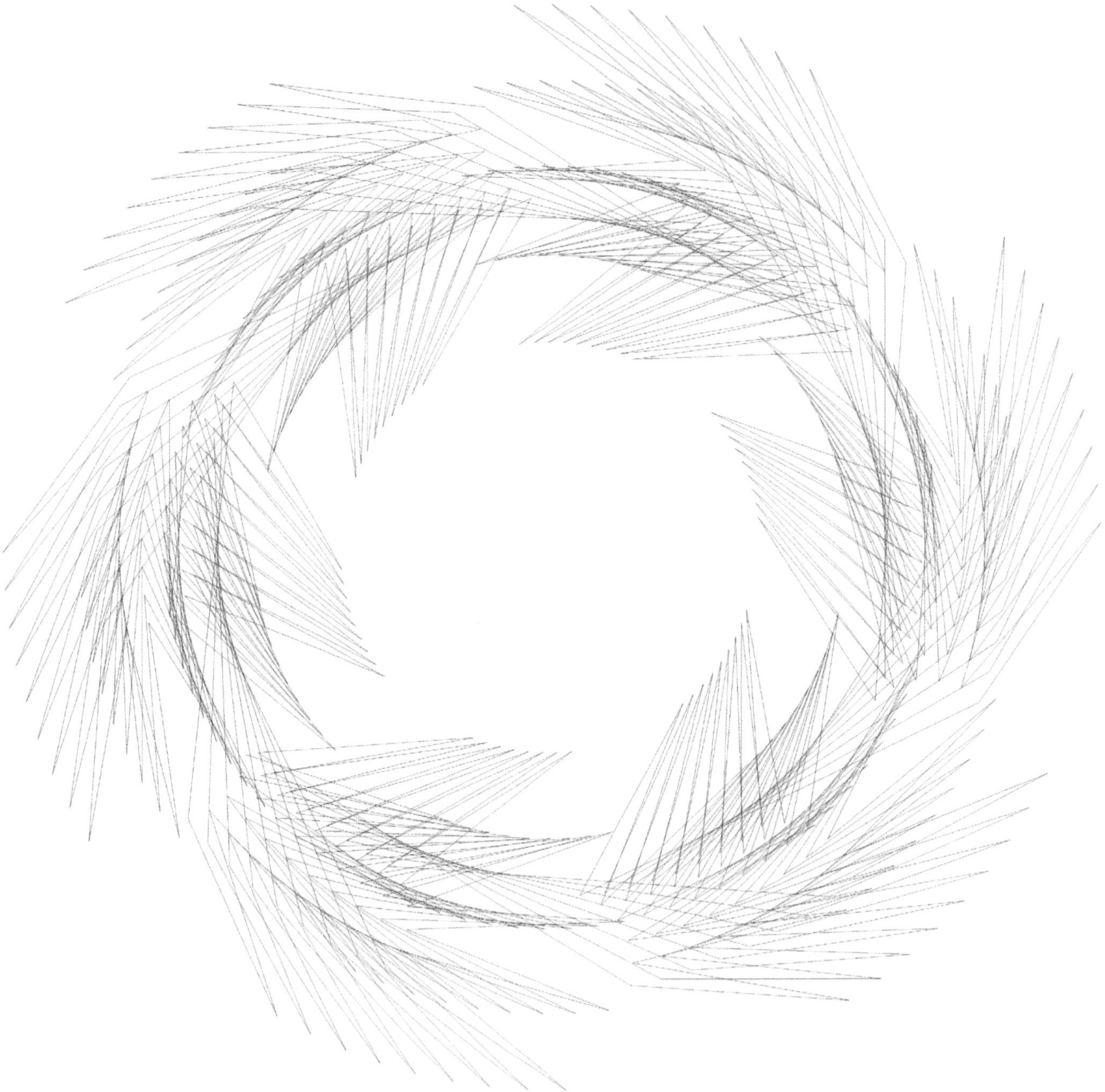

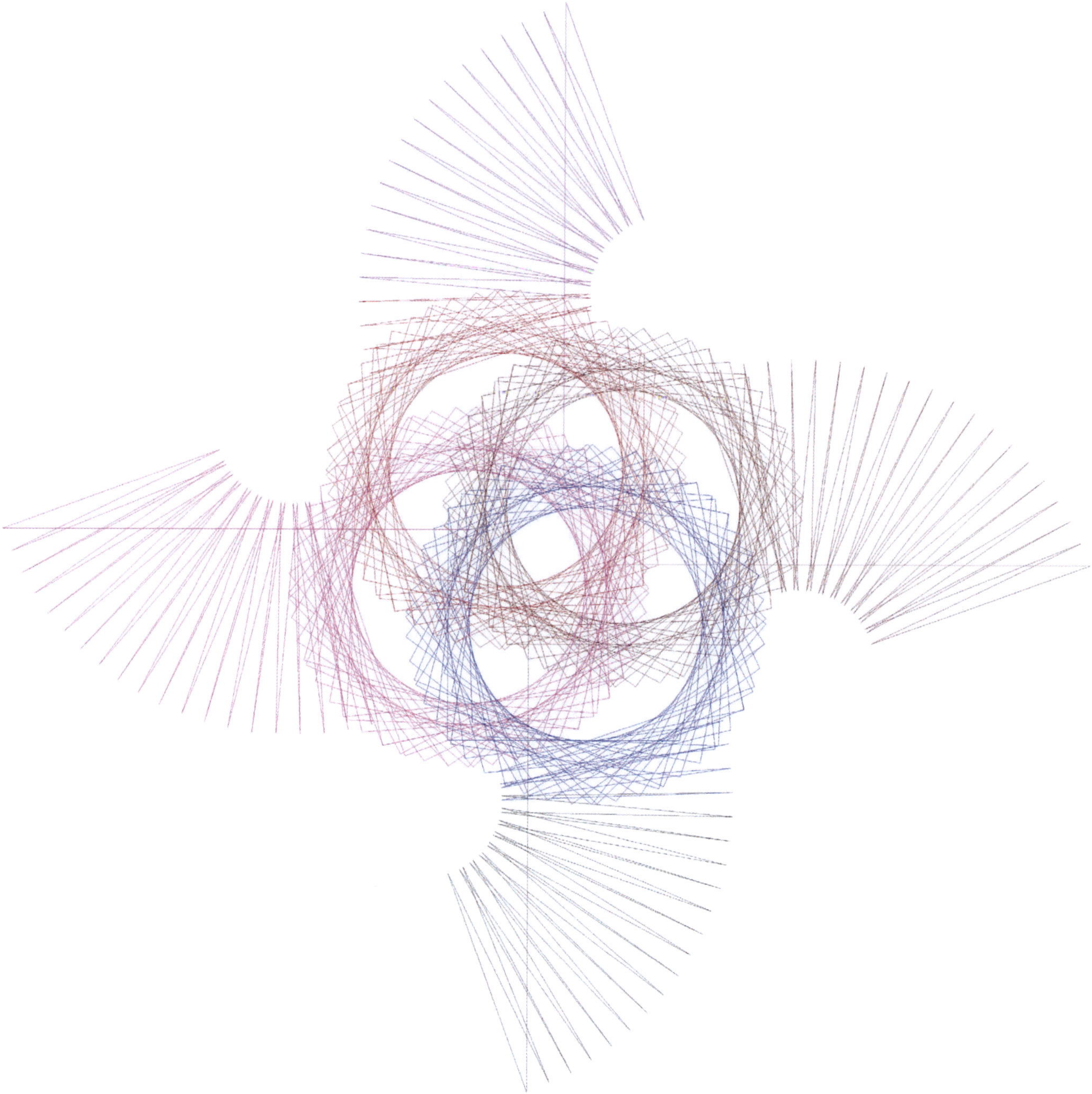

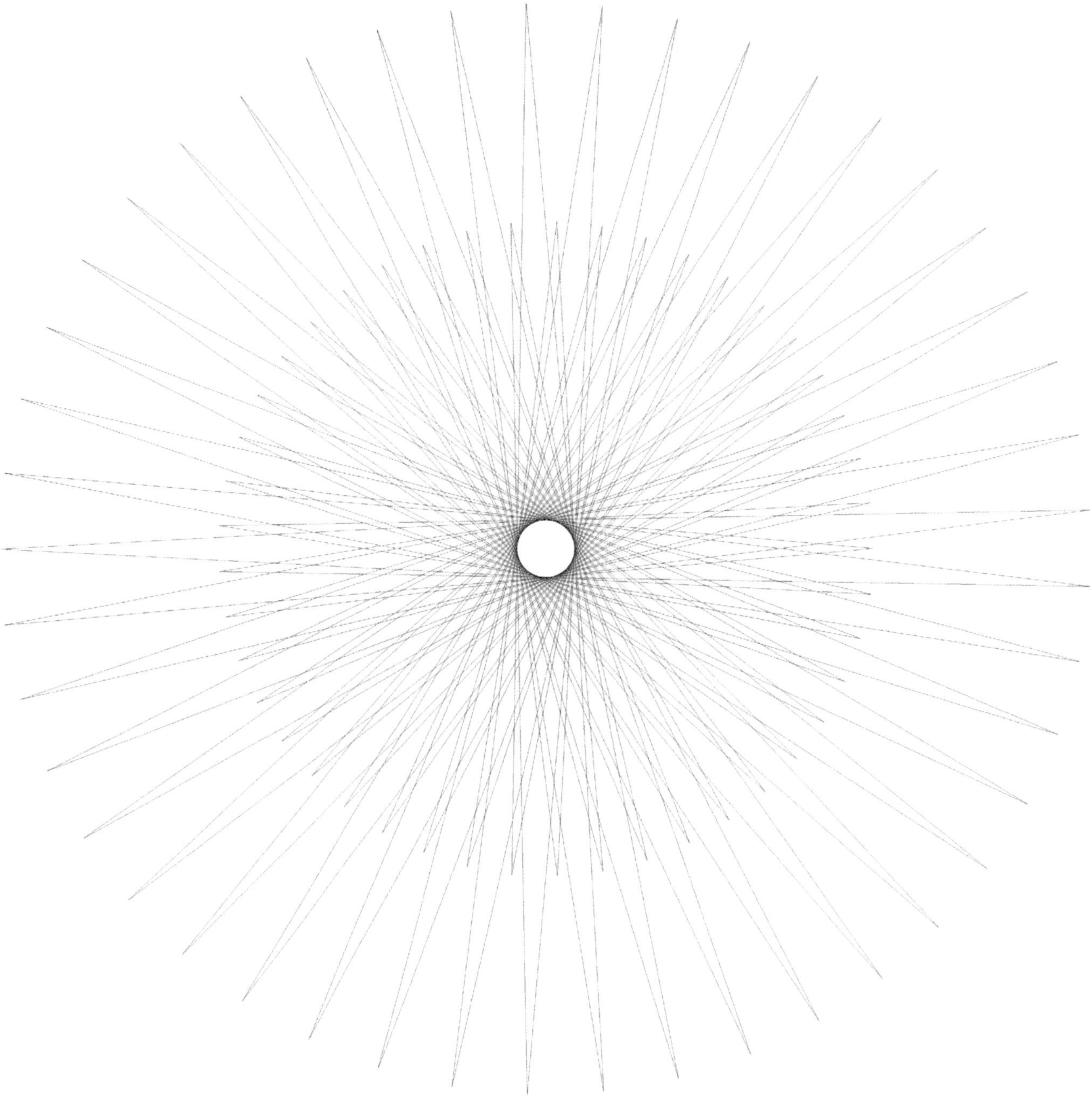

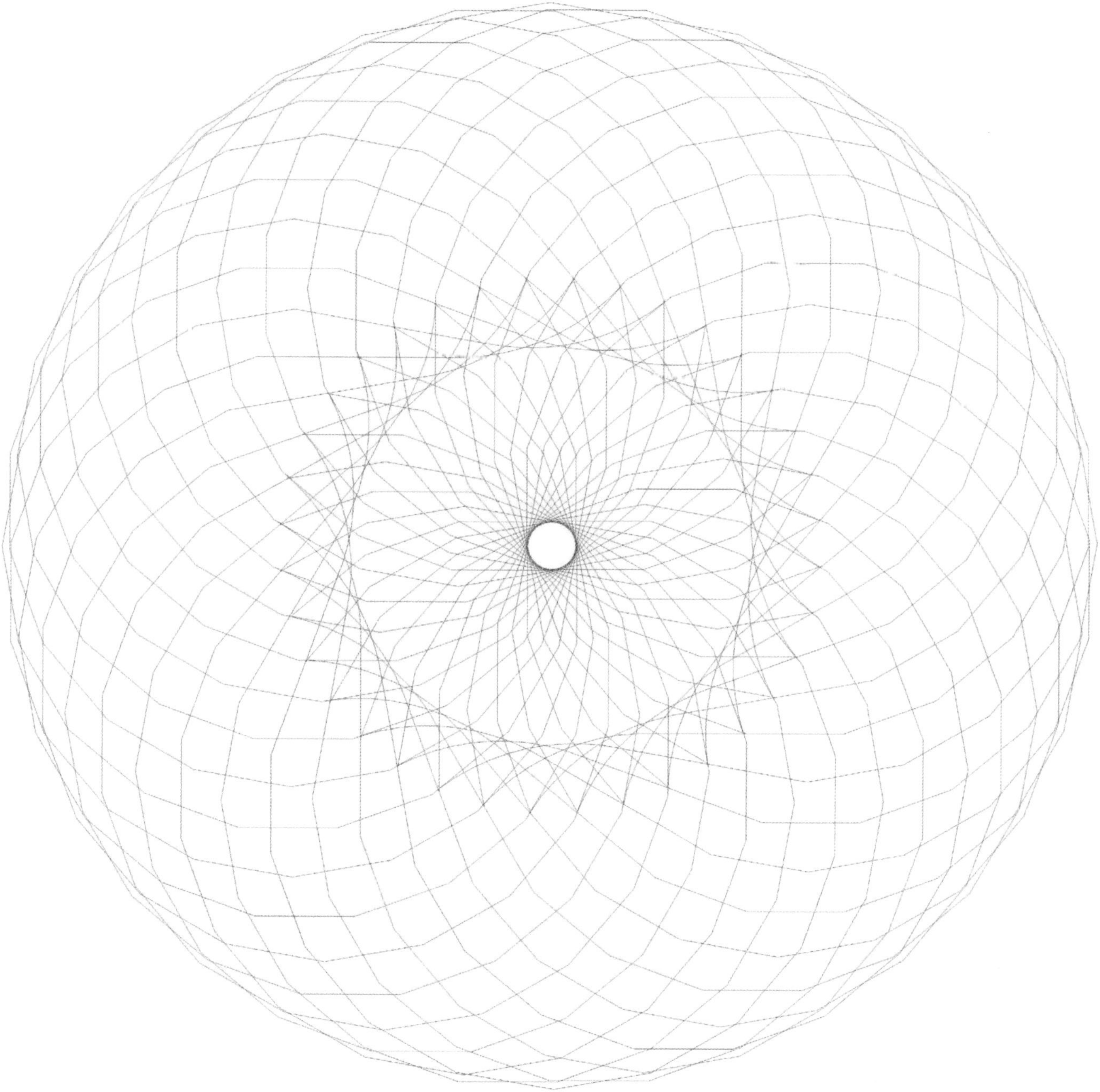

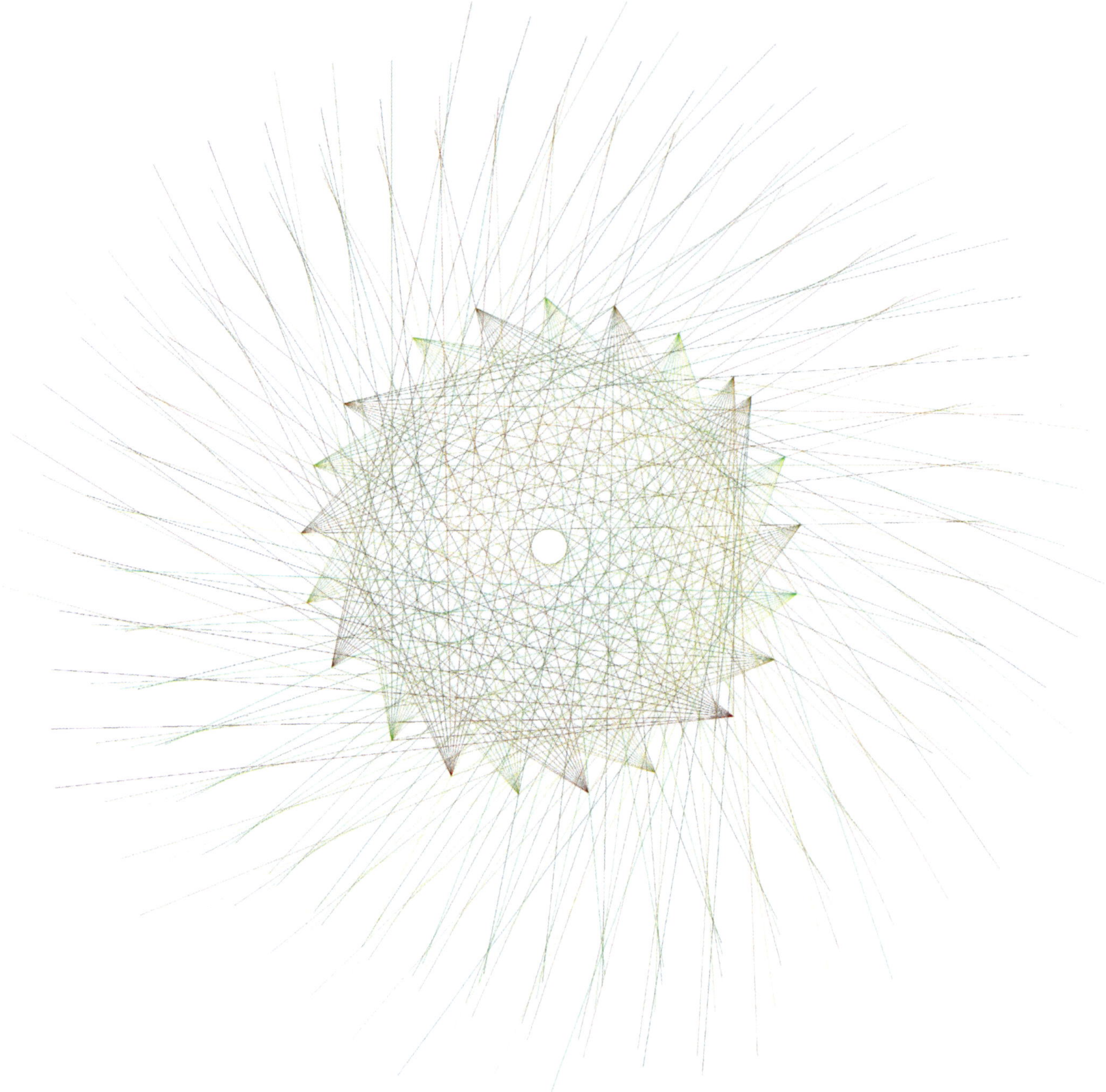

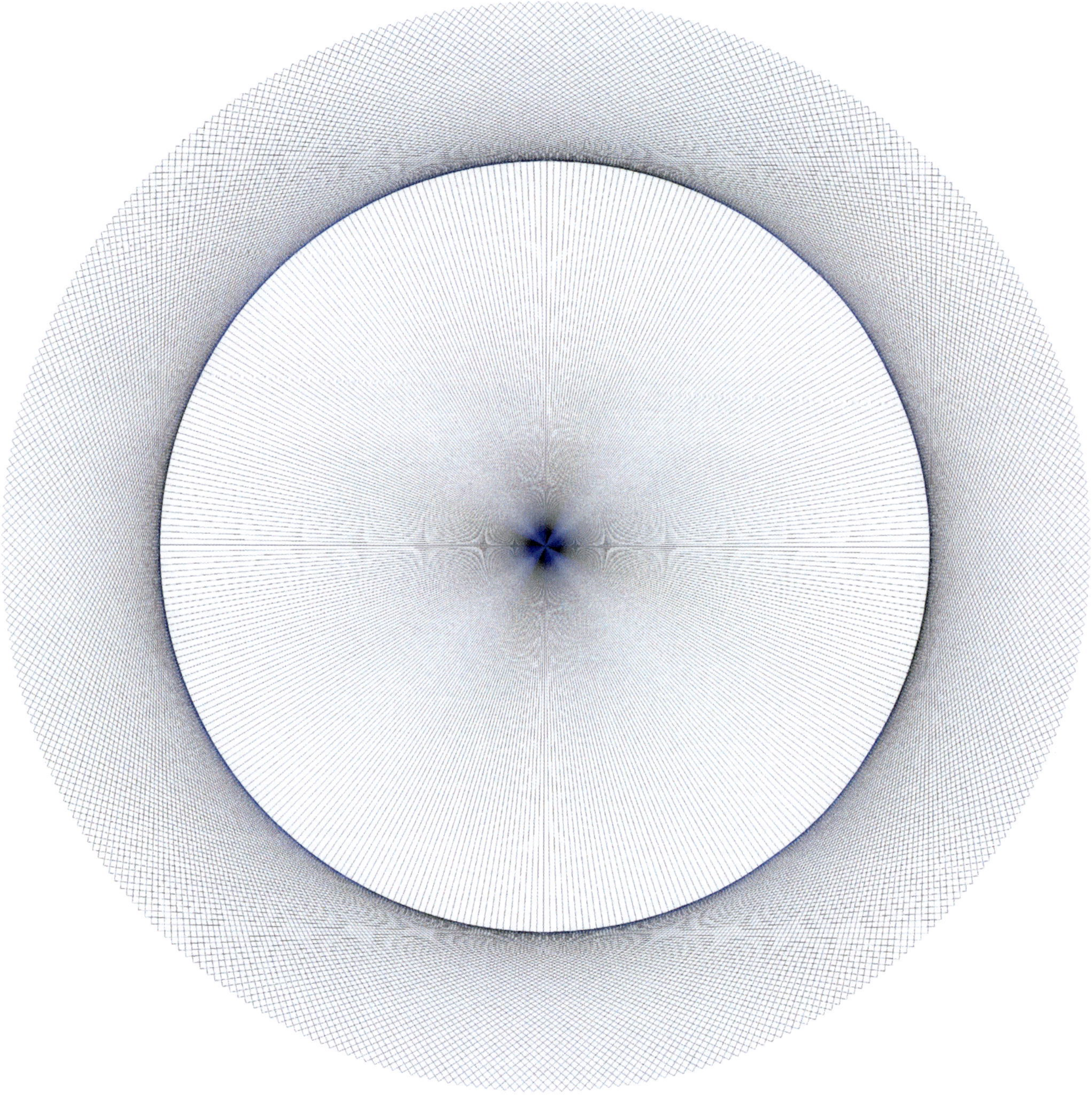

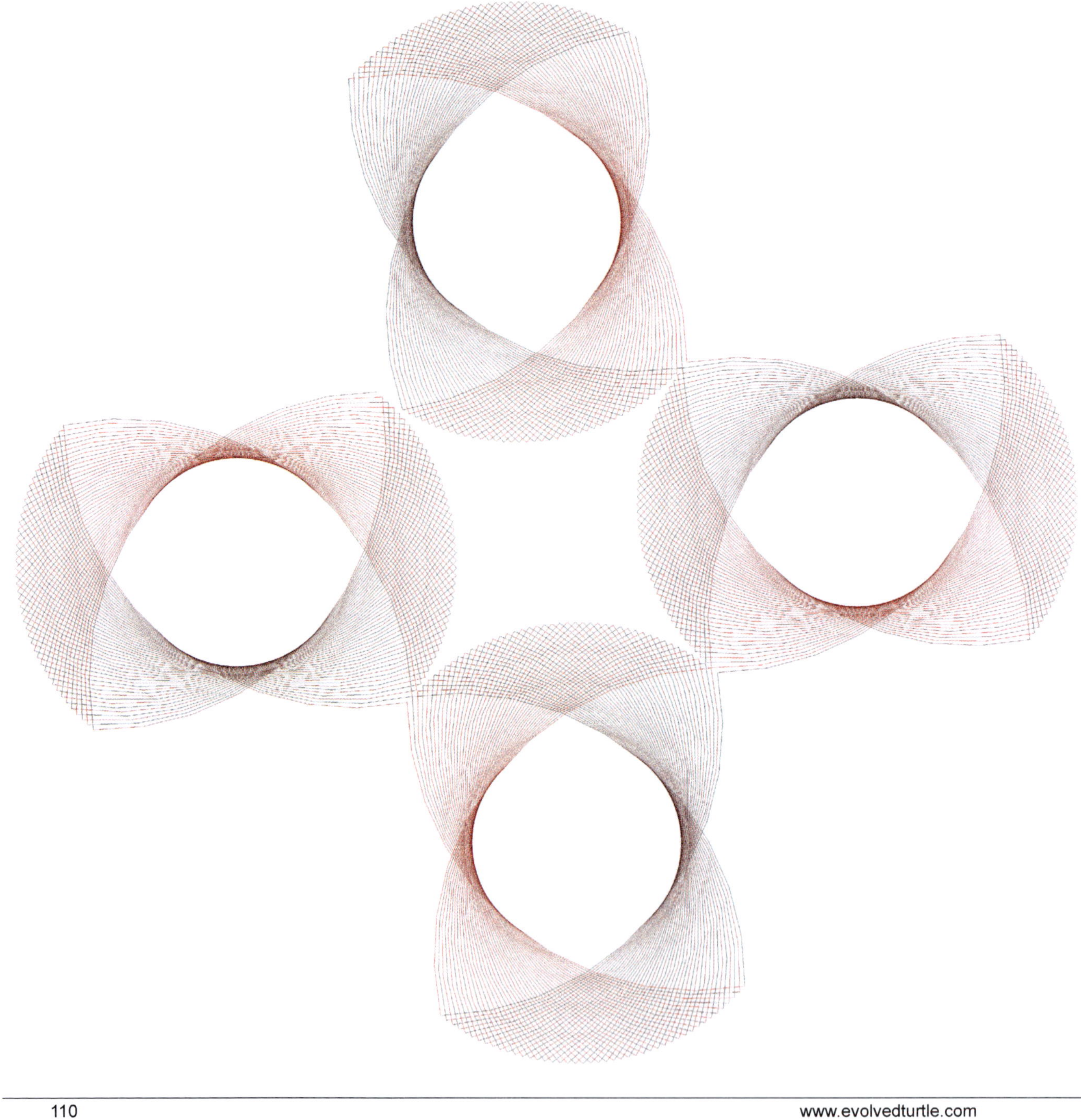

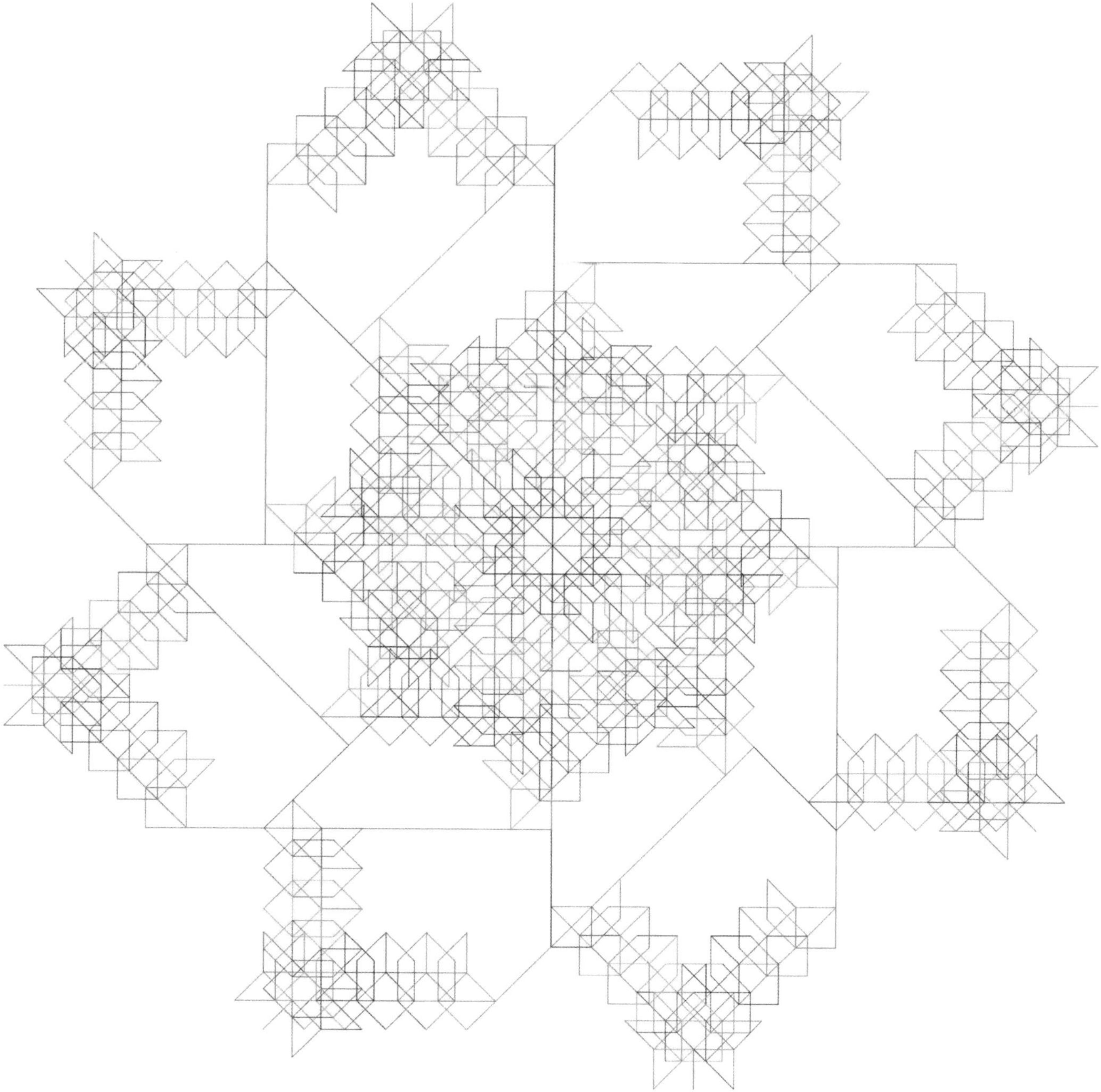

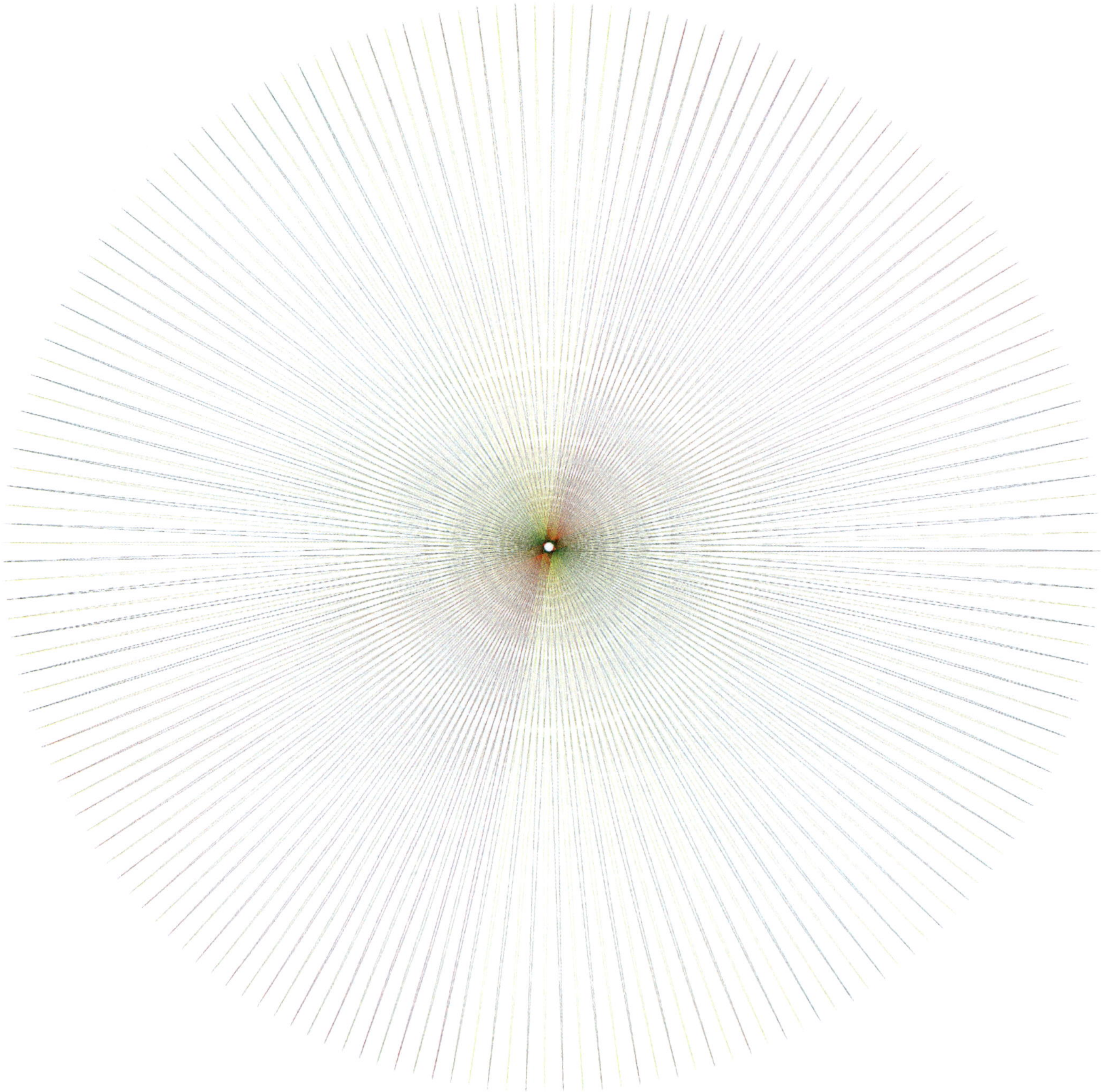

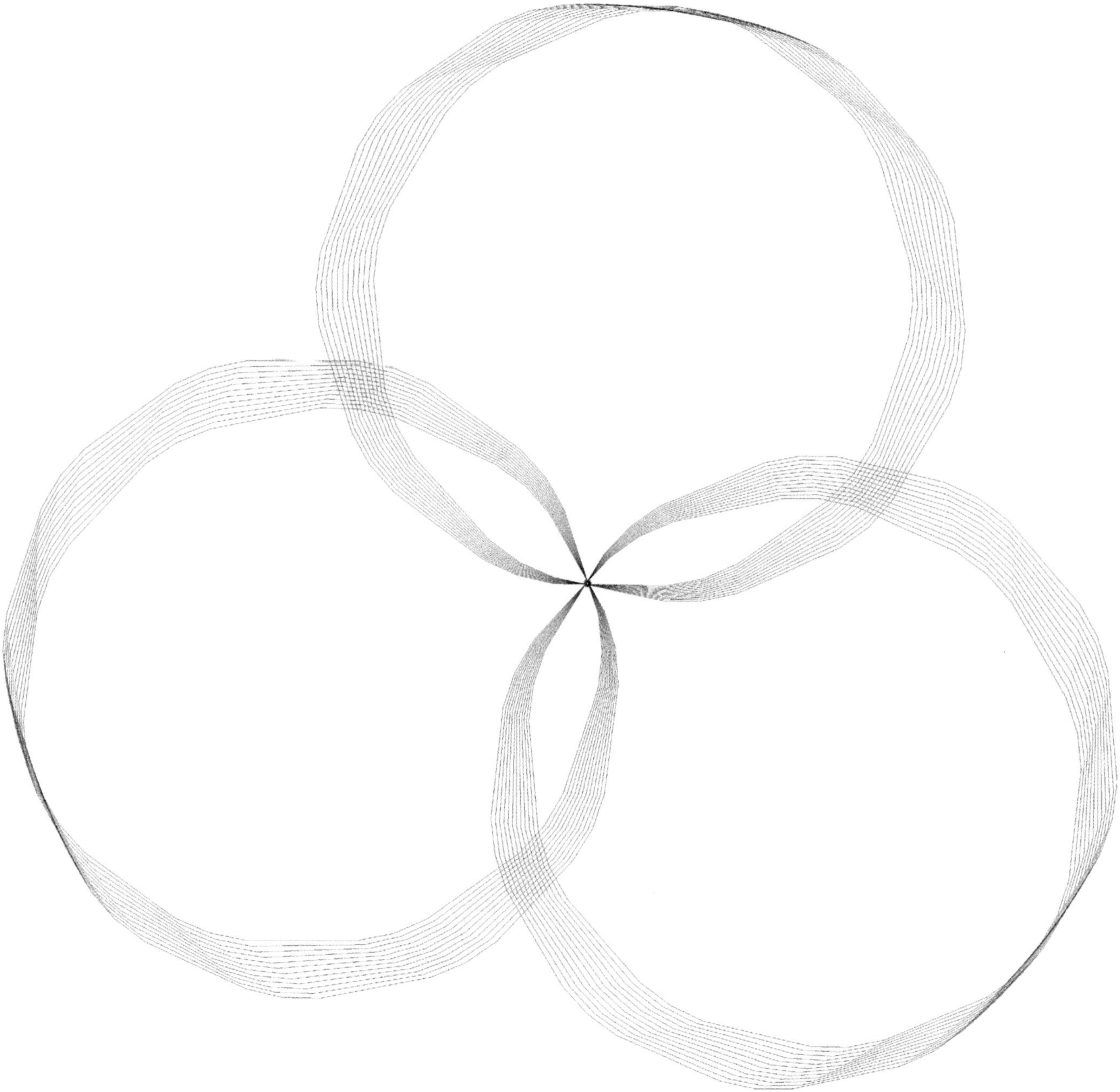

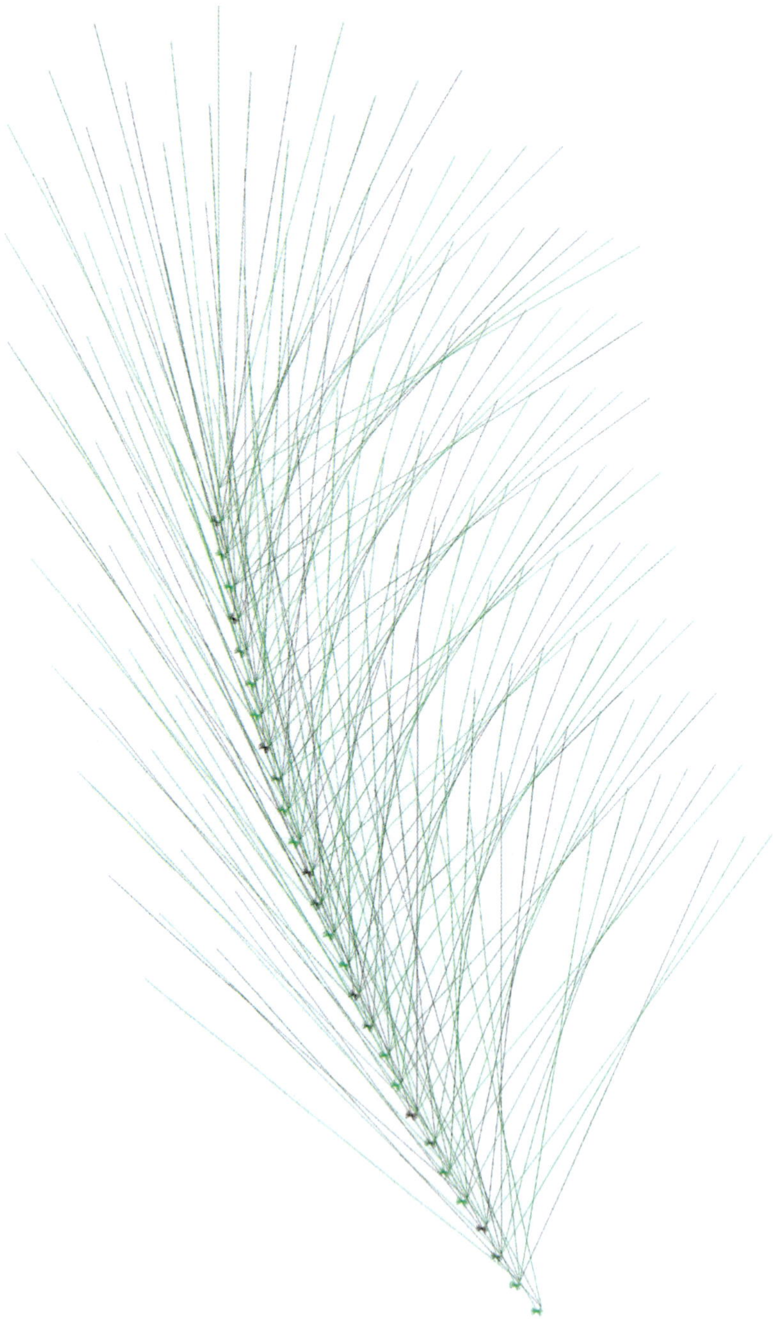

Chapter 5

Generations

I hope that you agree that turtles can be interesting. However, as a population engineer, you will also find interesting phenomena between generations of turtles. You will see heritage and diversity. You may even begin to collect families of turtles.

Often times you will see a turtle that is intriguing, but it has some imperfection that you find unacceptable. You may try to reduce the population to that single turtle and attempt to evolve away the imperfection that you dislike, but in the process create several turtles that are very similar because they share a lot of common code. Each turtle may be distinct in its own way, and yet you can tell that they are all sisters.

Other times you will have two turtles, each of which possesses some interesting characteristic, and you are attempting to cross them in the hopes that you can combine them into a single turtle. Maybe you are attempting to get a particular turquoise color into a mandala that you especially like.

The following pages present groups of turtles that are related to each other over generations. Each page presents four related turtles.

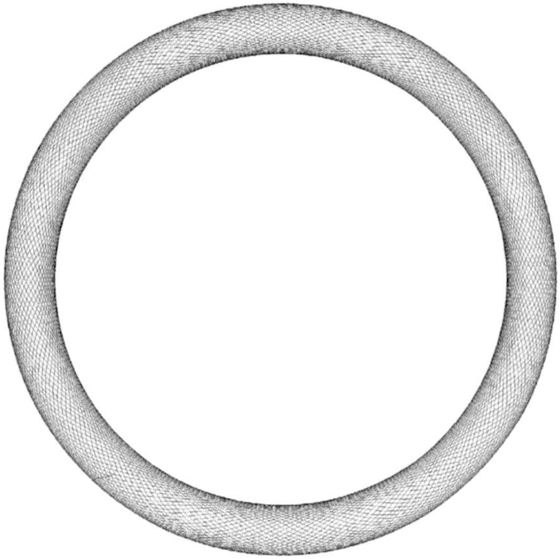

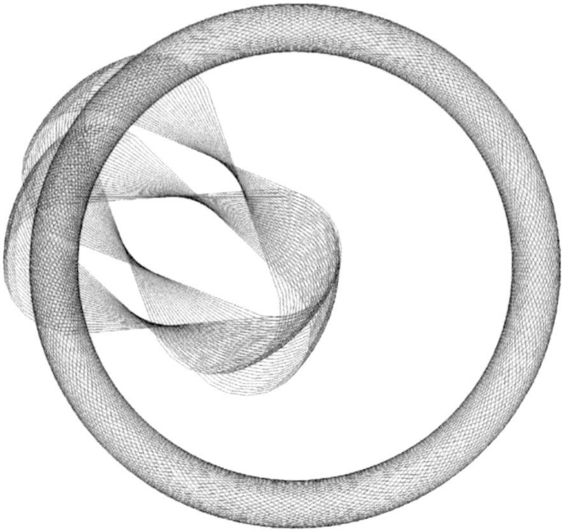

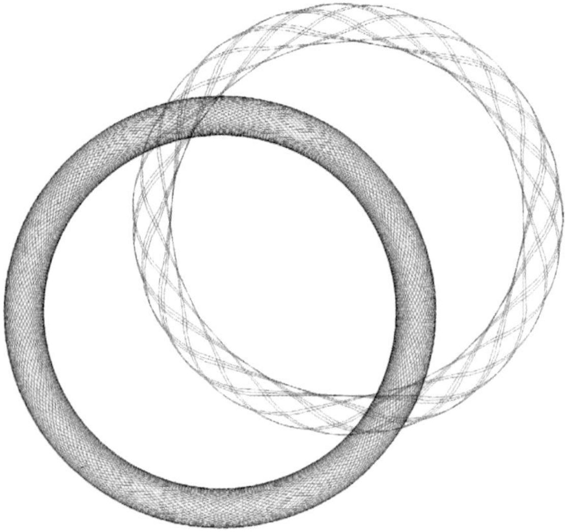

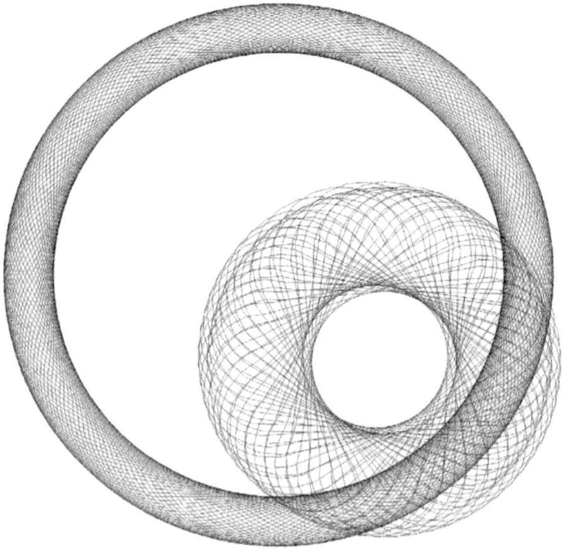

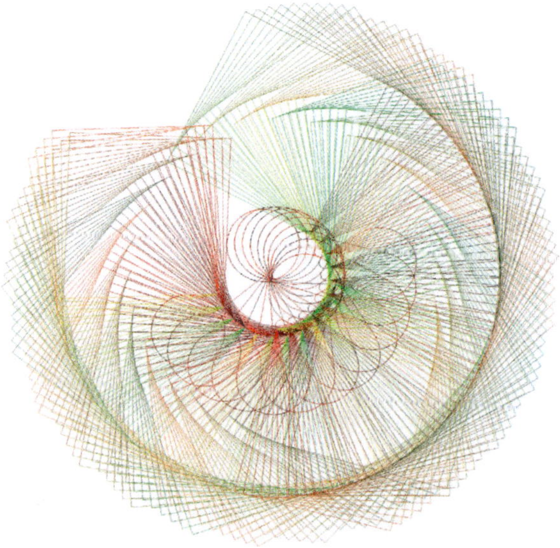

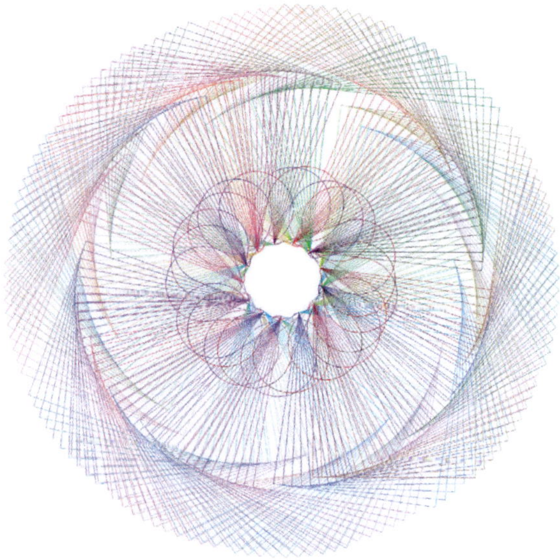

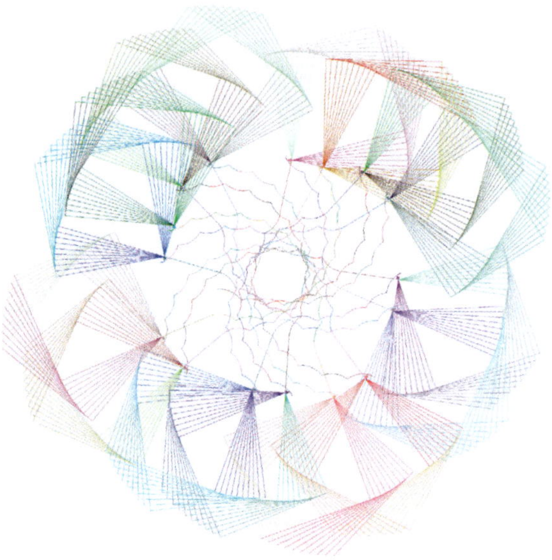

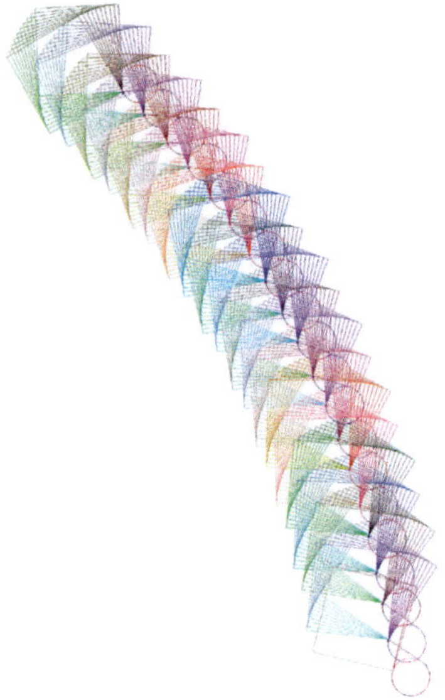

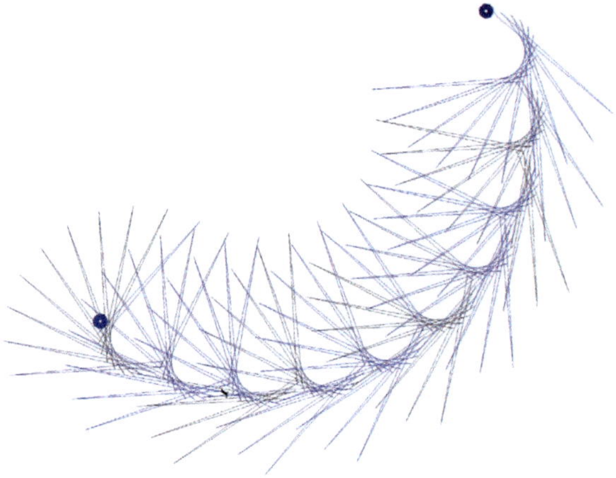

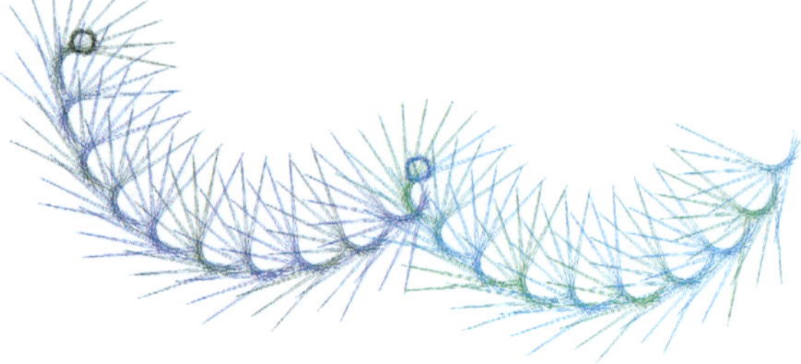

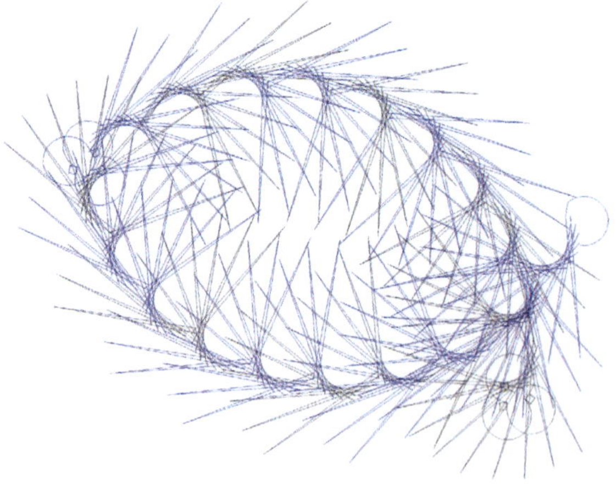

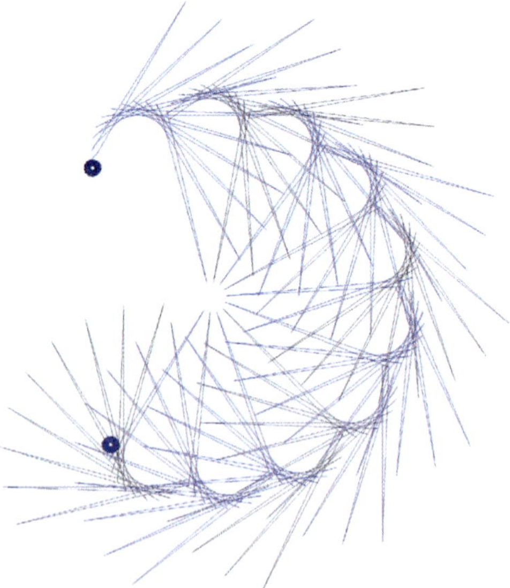

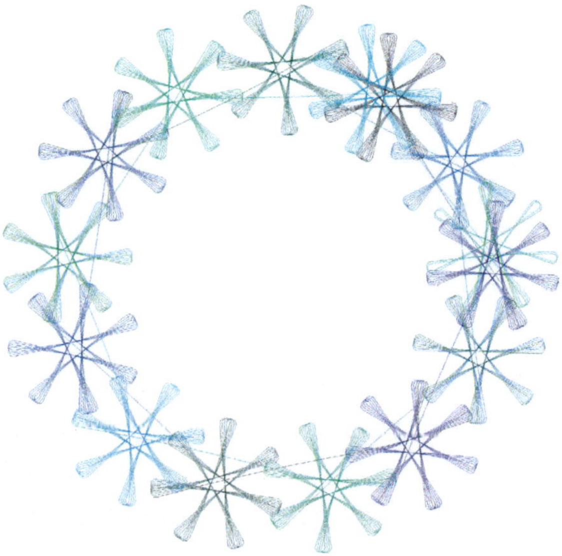

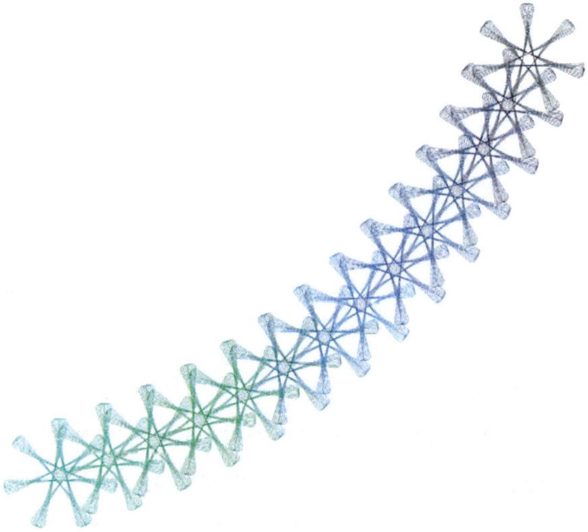

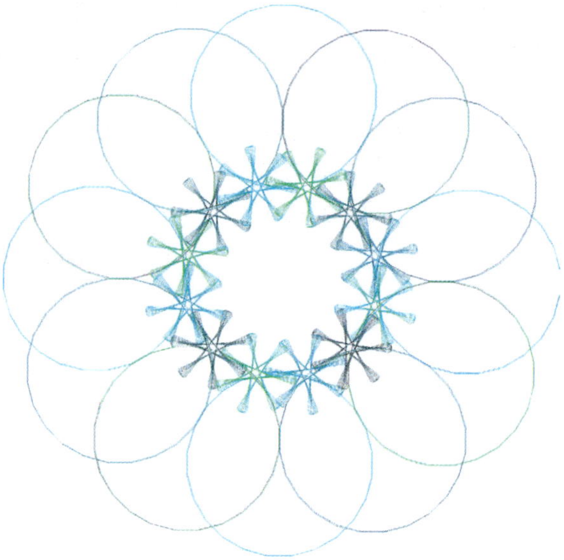

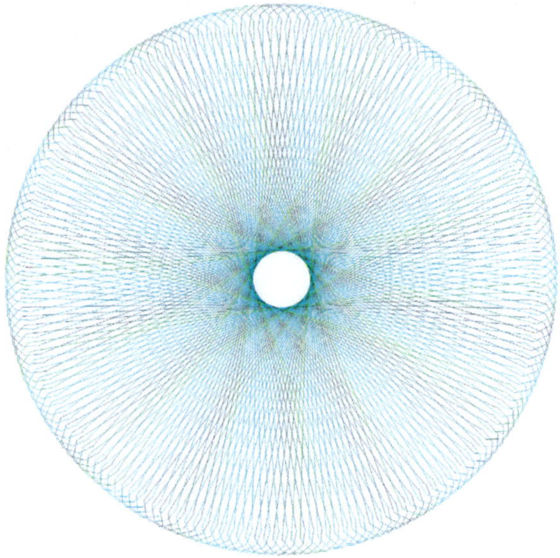

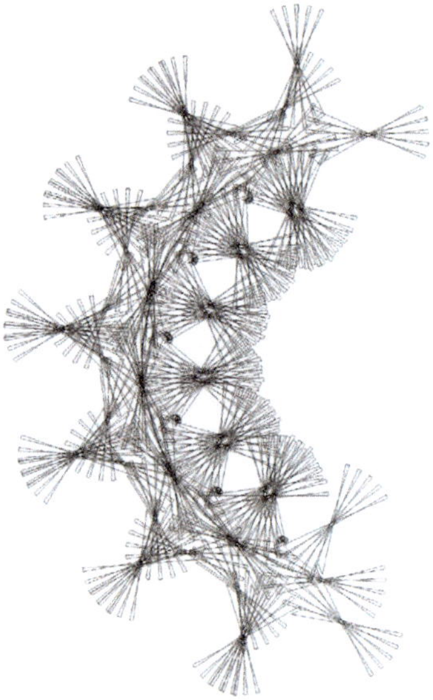

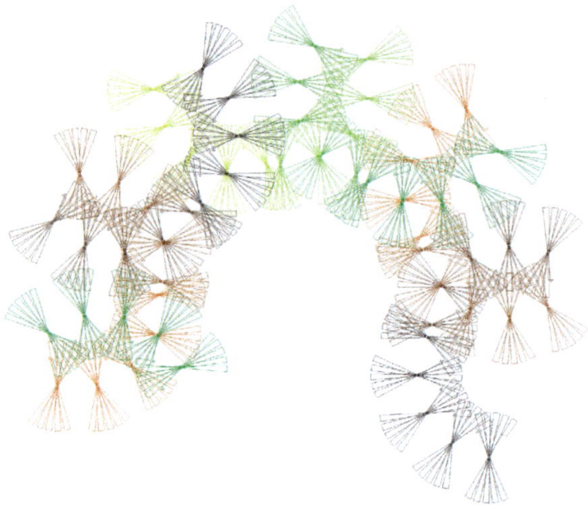

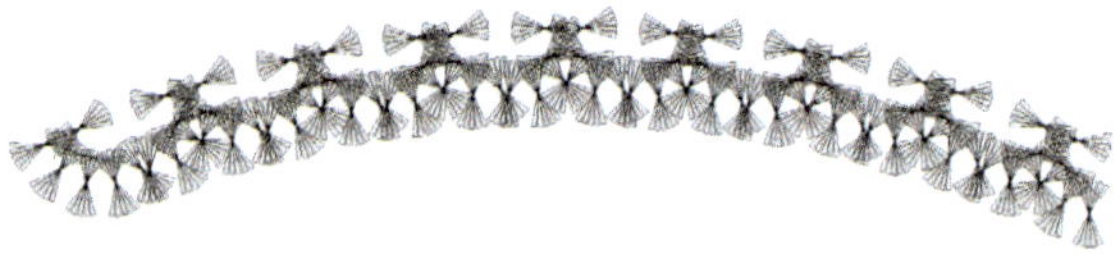

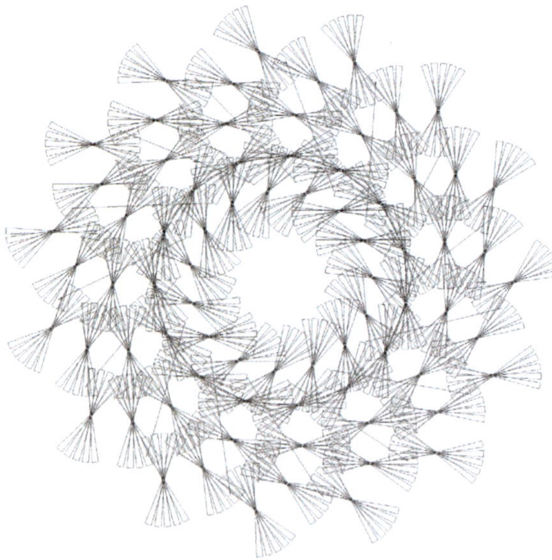

Chapter 6

Tiling

Occasionally a turtle will crop up that draws a familiar geometric shape, such as a square or triangle, or even a sphere or doughnut shape. The more interesting of these types of shapes are actually somewhat rare due to a number of aspects of the turtle drawing algorithm. These turtles find repetitive geometric outlines that create a mosaic pattern that one might find on a tile floor.

There is a great book on the subject of tiling and mosaics and their mathematical descriptions – *Tilings And Patterns*, by Grunbaum and Shepard. Fantastic book if you can locate a copy. The book explores the history, mathematics and algorithms of mosaic patterns and tiling.

The following pages present a few of what I call tiling turtles. As I stated, these turtles can be somewhat rare, so I have a limited number in my collection. It is an area worth pursuing, and I believe that they could be made a little more common if I were to spend time tweaking the fitness settings in BioLogo.

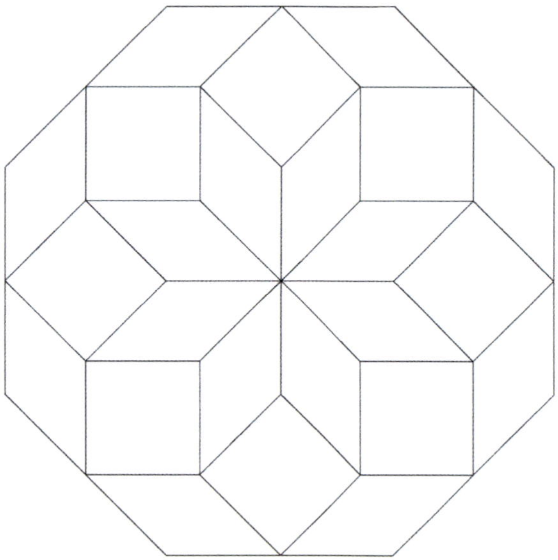

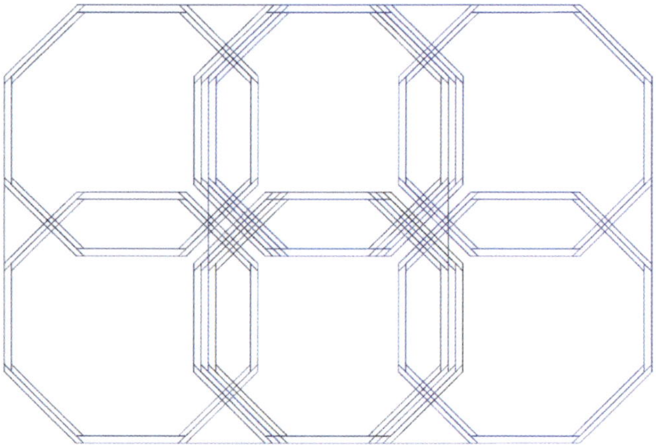

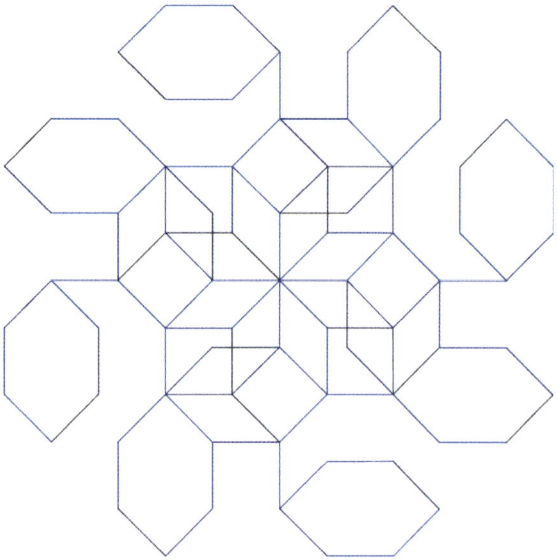

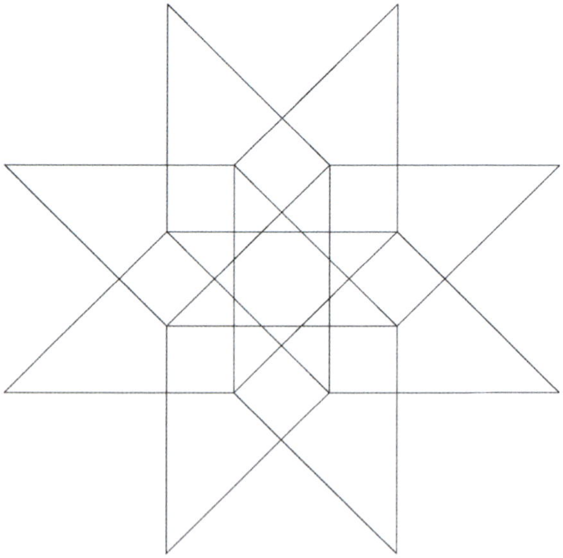

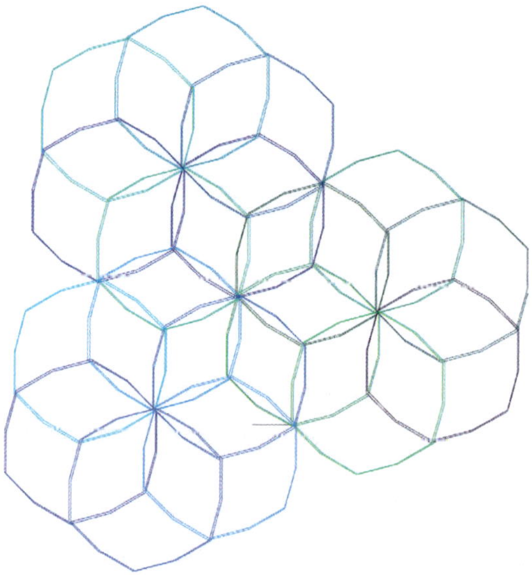

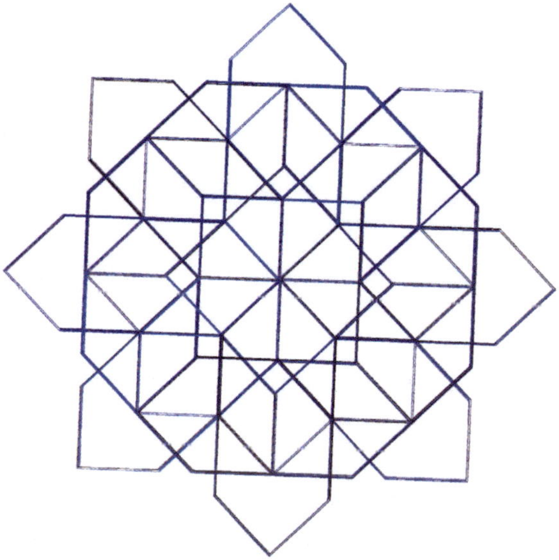

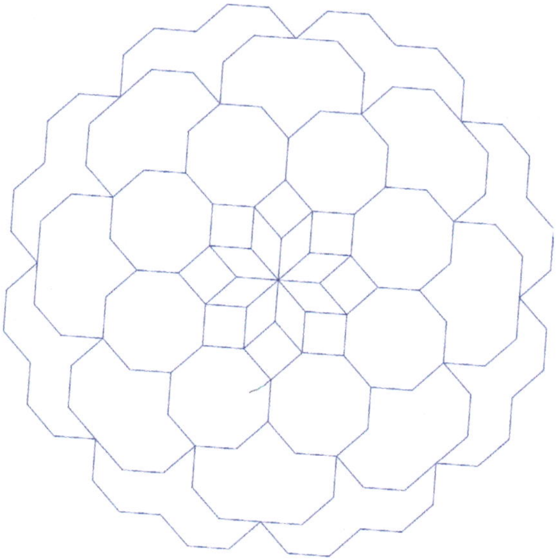

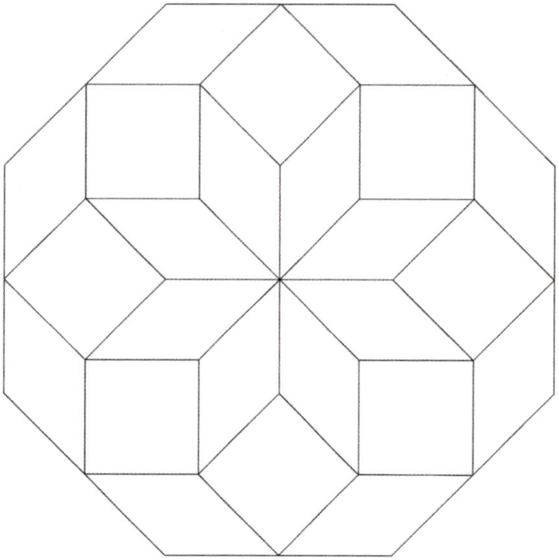

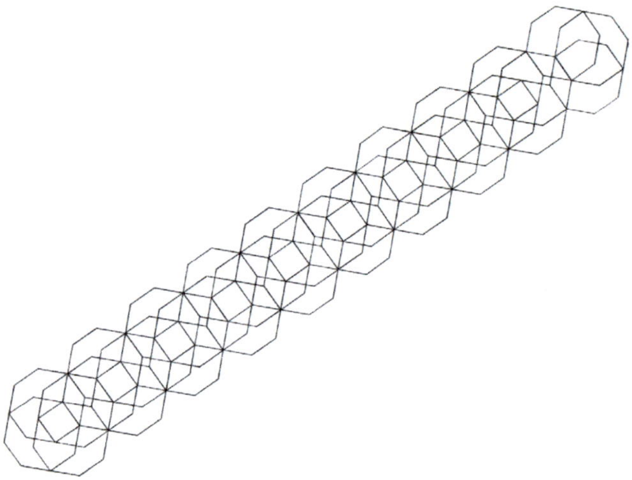

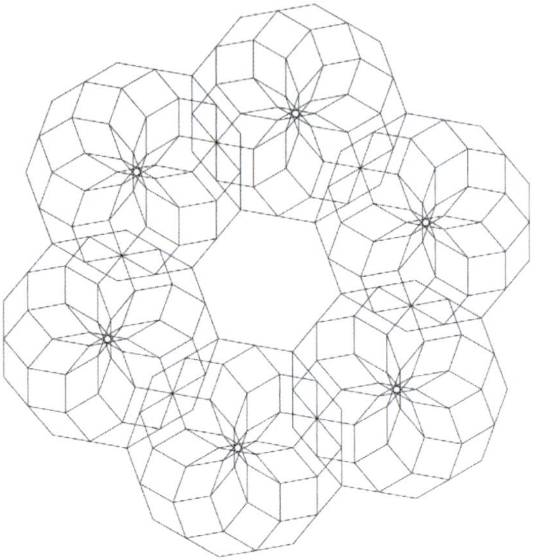

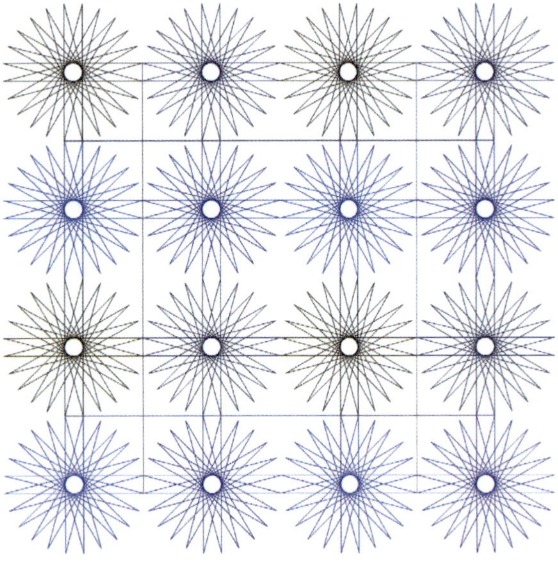

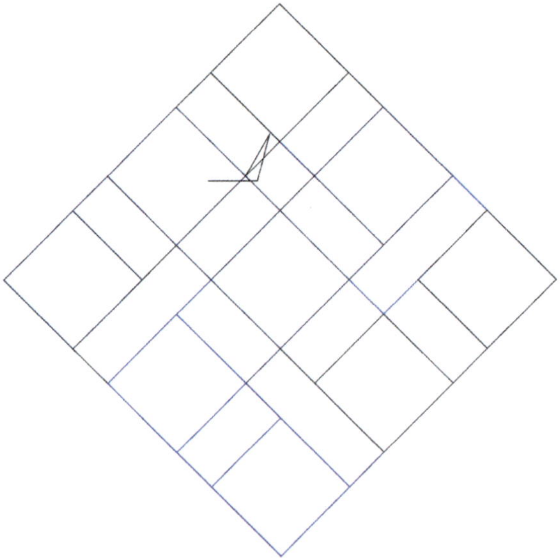

Chapter 7

Turtle Programs

For those who are a little more technically inclined or simply curious, here are a few turtles with their corresponding code.

Turtles are very simple-minded. They only know a few instructions. They can turn left, turn right, move forward, change their pen color, and raise and lower their pen. The turning, moving, and color change instructions have variations to specify distance or color or degrees of turn, as opposed to using a parameter. In other words, the turtles say "move forward 5", as opposed to the traditional "move forward N, where N = 5". For example, turtles have an instruction for `forward_1` and `forward_5`. If the turtle wants to move forward 6 units, it must execute two instructions: `forward_5`, and `forward_1`. Traditional Turtle Logo would use: `forward 6`.

The difference may seem subtle, but it is a critical feature of the language that greatly simplifies the implementation of the genetic programming engine. Otherwise, the splicing code would have to go through a great deal of effort to determine if two code sub-trees were compatible for crossover. In other words, we can combine forward and 6 together, but we cannot combine forward and forward, nor can we combine 6 and 6. By eliminating the concept of passing arguments to functions, we greatly simplified our code and the effort needed to implement it.

I apologize for the poor formatting of the code.

Purple Eyes

```
( REPEAT_10
    ( LEFT_3 )
    ( REPEAT_10
        ( REPEAT_5
            ( BLU_16 )
            ( BLOCK ( RED_3 ) ( FWD_50 ) ( LEFT_90 ) )
            ( LEFT_10 ) )
        ( RED_2 )
        ( REPEAT_10
            ( REPEAT_10
                ( REPEAT_5
                    ( BLOCK ( RED_3 ) ( FWD_50 ) ( LEFT_90 ) )
                    ( RIGHT_1 ) )
                ( RED_2 )
                ( REPEAT_3
                    ( BLOCK ( RED_3 ) ( FWD_50 ) ( LEFT_90 ) )
                    ( REPEAT_5 ( BLU_16 ) ( LEFT_10 ) ) ) )
            ( RED_2 ) ) ) )
```

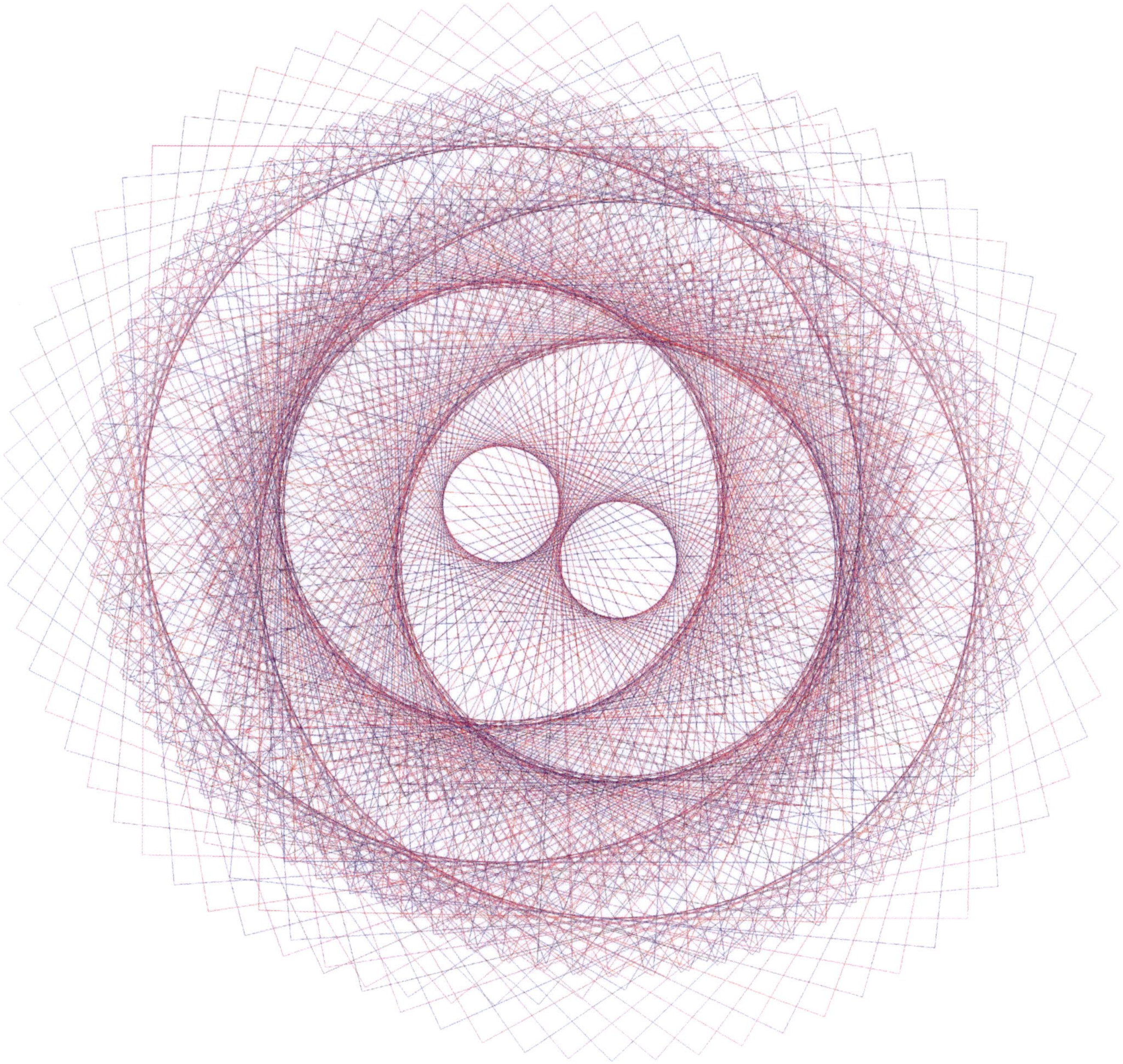

Spikey Ball

```
( REPEAT_10 ( REPEAT_5 ( REPEAT_3 ( IF_PEN_UP ( REPEAT_50 ) ( REPEAT_2 ( REPEAT_2
( BLOCK ( REPEAT_5 ( REPEAT_50 ) ( IF_PEN_UP ( BLOCK ( REPEAT_3 ( FWD_3 ) ( LEFT_1 ) )
( REPEAT_100 ) ( FWD_2 ) ) ( REPEAT_50 ) ) ( REPEAT_3 ( FWD_50 ) ( REPEAT_3 ( BLU_16 )
( FWD_5 ) ) ( BLOCK ( REPEAT_10 ( REPEAT_50 ) ( REPEAT_100 ) ( REPEAT_5 ( REPEAT_3
( REPEAT_100 ) ( REPEAT_3 ( BLU_16 ) ( REPEAT_5 ( FWD_100 ) ( RIGHT_90 ) ( RIGHT_90 )
( LEFT_5 ) ) ) ( REPEAT_5 ( FWD_100 ) ( RIGHT_90 ) ( RIGHT_90 ) ( LEFT_5 ) ) )
( REPEAT_3 ( IF_PEN_UP ( LEFT_3 ) ( LEFT_45 ) ) ( REPEAT_100 ) ) ) ) ( FWD_10 ) ) )
( REPEAT_50 ) ) ( REPEAT_50 ) ( REPEAT_100 ) ) ( REPEAT_100 ) ) ( REPEAT_5
( REPEAT_100 ) ( REPEAT_100 ) ( REPEAT_2 ( REPEAT_2 ( RIGHT_1 ) ( FWD_3 ) ( LEFT_1 ) )
( BLOCK ( REPEAT_100 ) ( REPEAT_5 ( IF_PEN_UP ( LEFT_10 ) ( RIGHT_2 ) ) ( BLOCK
( RIGHT_90 ) ( PEN_DOWN ) ( RIGHT_3 ) ) ( REPEAT_5 ( RIGHT_5 ) ( GRN_64 ) ( RIGHT_3 )
( GRN_1 ) ) ) ) ) ) ( REPEAT_50 ) ) ) ( REPEAT_50 ) ) ( LEFT_3 ) ) ( FWD_2 ) )
```

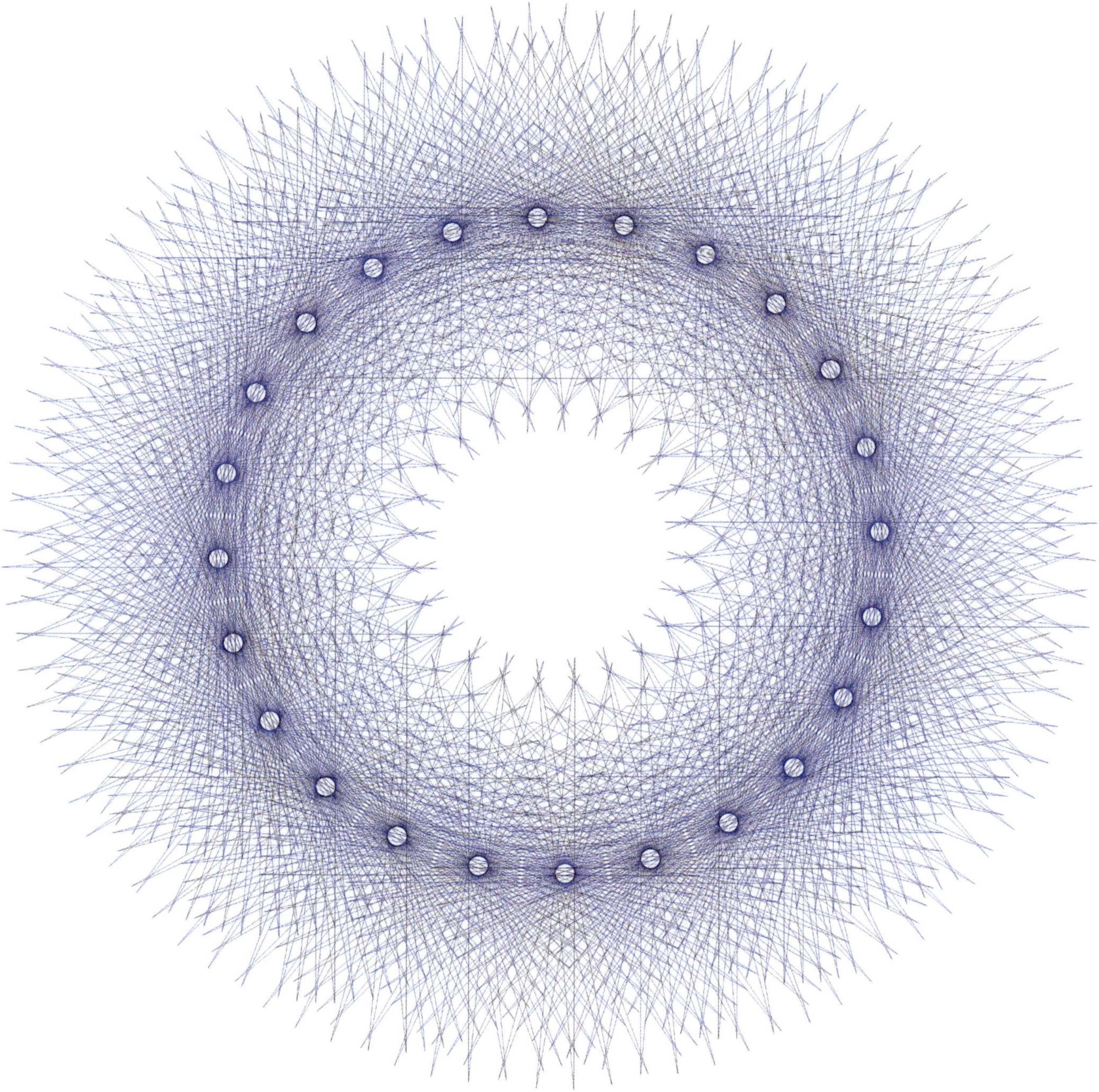

Son Of Mandala

```
( REPEAT_10 ( REPEAT_50 ) ( IF_PEN_UP ( RIGHT_45 ) ( REPEAT_50 ) )
( REPEAT_50 ) ( REPEAT_5 ( REPEAT_5 ( REPEAT_5 ( RIGHT_1 ) ( IF_PEN_UP
( IF_PEN_DOWN ( FWD_10 ) ( RIGHT_15 ) ) ( REPEAT_2 ( LEFT_90 ) ( LEFT_10 )
( FWD_10 ) ) ) ( RIGHT_5 ) ( FWD_100 ) ) ( REPEAT_5 ( RIGHT_1 ) ( IF_PEN_UP (
IF_PEN_DOWN ( FWD_10 ) ( RIGHT_15 ) ) ( REPEAT_2 ( LEFT_90 ) ( LEFT_10 )
( FWD_10 ) ) ) ( RIGHT_5 ) ( FWD_100 ) ) ) ( REPEAT_3 ( REPEAT_2 ( FWD_5 )
( RIGHT_10 ) ) ( BLOCK ( LEFT_2 ) ( RIGHT_2 ) ( FWD_5 ) ( LEFT_15 ) )
( REPEAT_50 ) ) ( IF_PEN_UP ( IF_PEN_UP ( REPEAT_50 ) ( REPEAT_10 ( PEN_UP )
( RIGHT_1 ) ( FWD_50 ) ( LEFT_90 ) ) ) ( IF_PEN_UP ( GRN_1 ) ( LEFT_10 ) ) )
( IF_PEN_UP ( IF_PEN_DOWN ( FWD_10 ) ( RIGHT_15 ) ) ( REPEAT_2 ( LEFT_90 )
( LEFT_10 ) ( FWD_10 ) ) ) ) ) )
```

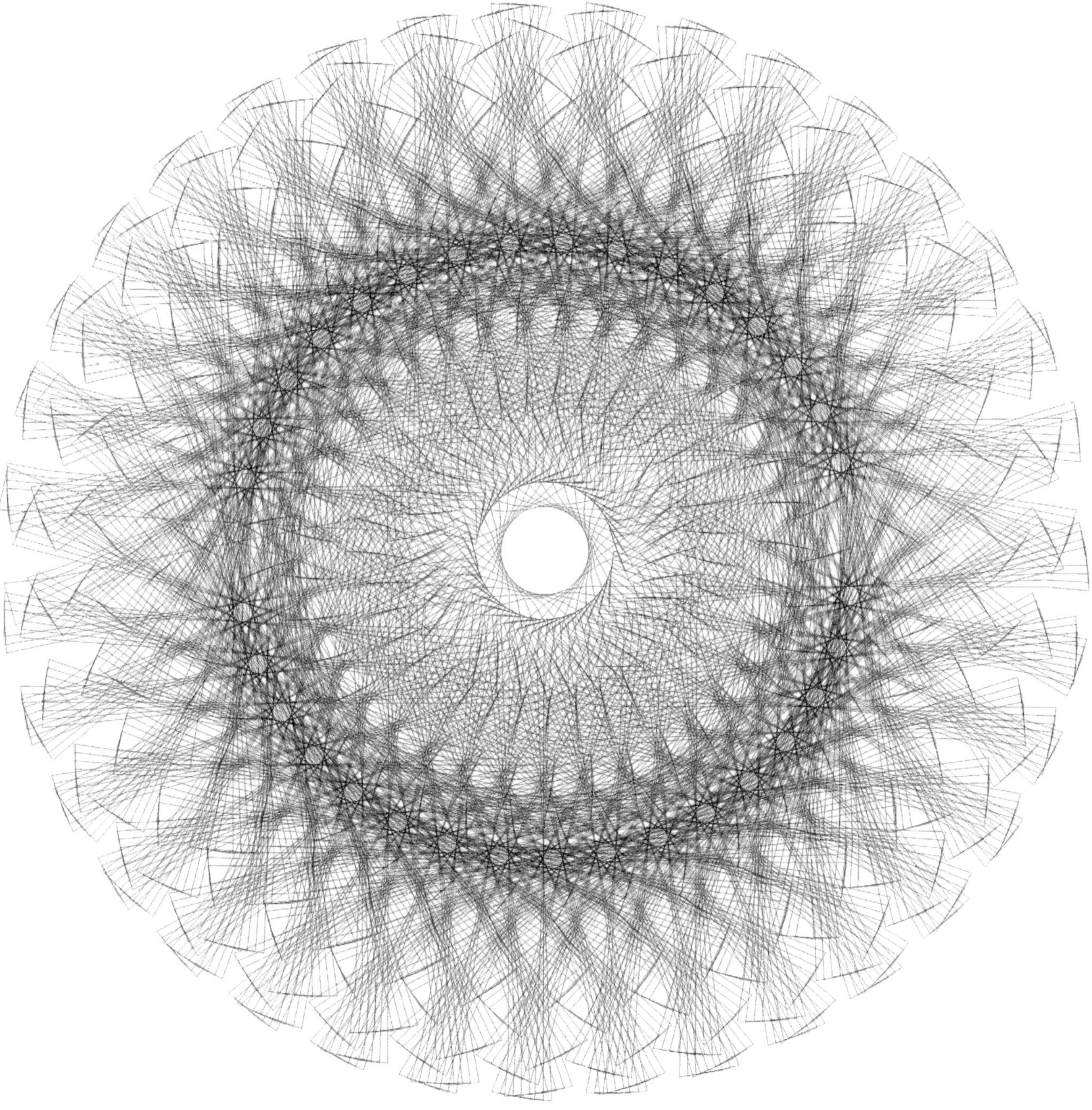

References

Websites

Evolved Turtle: www.evolvedturtle.com

Turtle Logo Wikipedia: www.wikipedia.org/wiki/Logo_(programming_language)

Books

Adaptation In Natural And Artificial Systems, By John H. Holland,
Published by The MIT Press
ISBN 0-2625-8111-6

Genetic Programming, By John R. Koza,
Published by The MIT Press
ISBN 0-2621-1170-5

Tilings And Patterns, By Branko Grunbaum and G. C. Shepard,
Published by W. H. Freeman And Company, New York
ISBN 0-7167-1193-1

Turtle Talk: A Beginner's Book Of Logo, By Seymour Simon,
Published by Thomas Y. Crowell & Company

Printed in Dunstable, United Kingdom

91218902R00080